T0332886

Stardust
from
Meteorites
An Introduction to
Presolar Grains

WORLD SCIENTIFIC SERIES IN ASTRONOMY AND ASTROPHYSICS

Editor: Jayant V. Narlikar
Inter-University Centre for Astronomy and Astrophysics, Pune, India

---

Published:

Volume 1:   Lectures on Cosmology and Action at a Distance Electrodynamics
              F. Hoyle and J. V. Narlikar

Volume 2:   Physics of Comets (2nd Ed.)
              K. S. Krishna Swamy

Volume 3:   Catastrophes and Comets*
              V. Clube and B. Napier

Volume 4:   From Black Clouds to Black Holes (2nd Ed.)
              J. V. Narlikar

Volume 5:   Solar and Interplanetary Disturbances
              S. K. Alurkar

Volume 6:   Fundamentals of Solar Astronomy
              A. Bhatnagar and W. Livingston

Volume 7:   Dust in the Universe: Similarities and Differences
              K. S. Krishna Swamy

Volume 8:   An Invitation to Astrophysics
              T. Padmanabhan

Volume 9:   Stardust from Meteorites: An Introduction to Presolar Grains
              M. Lugaro

Volume 10:   Rotation and Accretion Powered Pulsars
              P. Ghosh

Volume 11:   Find a Hotter Place!: A History of Nuclear Astrophysics
              L. M. Celnikier

Volume 12:   Physics of Comets (3rd Edition)
              K. S. Krishna Swamy

Volume 13:   From Black Clouds to Black Holes (3rd Edition)
              J. V. Narlikar

*Publication cancelled.

Stardust
from
Meteorites

An Introduction to
Presolar Grains

Maria Lugaro
Astronomical Institute, Utrecht University
The Netherlands

World Scientific

NEW JERSEY · LONDON · SINGAPORE · BEIJING · SHANGHAI · HONG KONG · TAIPEI · CHENNAI

Published by

World Scientific Publishing Co. Pte. Ltd.

5 Toh Tuck Link, Singapore 596224

USA office: 27 Warren Street, Suite 401-402, Hackensack, NJ 07601

UK office: 57 Shelton Street, Covent Garden, London WC2H 9HE

British Library Cataloguing-in-Publication Data
A catalogue record for this book is available from the British Library.

STARDUST FROM METEORITES
An Introduction to Presolar Grains

ISBN-13 978-981-256-099-5
ISBN-10 981-256-099-8

Printed in Singapore

to Brett

Preface

Holding a tiny piece of a star in one's hand might seem like an impossible dream. Today, however, it is a reality. Microscopic dust grains manufactured around stars that existed before our solar system formed, are now extracted from meteorites that have fallen to the Earth, and analysed in terrestrial laboratories. The stellar origin of this dust is revealed by its exotic composition, much different from that of the bulk of the material in the solar system.

Although the analysis of stellar grains is a very young field – it was born in 1987 – much has already been achieved, especially since recent advances in laboratory techniques have allowed to perform analysis of grains of size smaller than a micrometer, and high-precision measurements of the composition of elements present in trace in the grains. Soon, it will be possible to collect data on the composition of many different elements in single dust grains of many different types. In some instances, the technological advances have been so rapid that the theoretical interpretation can barely keep up with the flow of new information.

The aim of this book is to present issues related to stellar grains in an accessible way, thus helping students and scientists at all levels and of all backgrounds to learn about this field. Indeed, a broad awareness about stellar grain research and its implications is still lacking in the astronomy community at large, mainly because the subject is so new, and different when compared to anything that has been studied before. Both researchers and students need a broad basic knowledge and a clear presentation of the tools needed to familiarise themselves with presolar grains.

In spite of the vast amount of information that stellar grains yield about the different sites and processes that affect their features during their life – from the formation of dust around stars, to the survival of dust in the inter-

stellar medium, to the formation of our own solar system – the focus here is
on using the grains as evidence of the processes related to the evolution of
stars and to the nuclear reactions that change their composition throughout their lives (*nucleosynthesis*). In this respect, when confronted with the
data coming from grain analysis, the problem is two-fold: on one hand
we want to satisfy our immediate curiosity of answering the grain puzzle:
How did their unusual compositions come about? On the other hand, we
want to use the data as a tool in the wider task of understanding stars and
explaining how they have been producing the elements that constitute the
Universe, including ourselves and our environment. For this task, stellar
grains represent a breakthrough. They have recorded the composition of
stars, and their analysis yields data of extremely high-precision: error bars
are as low as 1%, as compared to the spectroscopic observations of the composition of stars, which are affected by uncertainties of approximately 50%.
Paraphrasing Clayton & Nittler [68], two- or three-dimensional diagrams
representing the composition of stellar grains could have a similar use for
the classification of nucleosynthesis processes, as the Hertzprung-Russell
diagram has for the classification of stars.

Even confining the discussion of this volume to the impact of the analysis
of stellar grains on studies of stellar evolution and nucleosynthesis, it has
not been feasible to cover in detail all of the many issues. Instead, the
focus is on one particular type of stellar grain, silicon carbide (SiC), since
these grains have been the most extensively analysed to date, and on one
particular type of stellar source, Asymptotic Giant Branch stars, which
appear to be the producers of many of the stellar grains recovered to date.
This focus stems from my own expertise in the subject and from a wish
to offer the reader detailed examples, displaying the level of sophistication
of the implications of the study of presolar grains. Other types of grains
and stellar sites would also be extremely interesting to discuss in detail,
for example grains that are believed to have originated from supernova
explosions. A large extension of the book would be necessary to treat these
other topics properly. In any case, an introduction and references are given
for all the issues that are not discussed in detail.

The future of stellar grain studies is dynamic and exciting: many puzzles
are far from being solved and many more will come to challenge us.

The book is divided into six chapters, of which the first three provide an
overview of the topic and of related basic information. The following three
chapters are, instead, more specific and thus represent a more complex
reading. They can be used to deepen the understanding of the origin of the

different types of grains, and of the different types of information that it is possible to extract in relation to nucleosynthesis processes in stars. Each chapter is equipped with a set of exercises, of which detailed solutions are given in Appendix B. Appendix A provides the reader with a general simple glossary, which should be useful to consult to clarify a word or expression, while reading the main text, or to remind oneself of their meaning. A set of selected references are given in Appendix C.

I thank Ernst Zinner, Roberto Gallino and John Lattanzio for supporting me and inspiring my work with their dedication to the study of presolar grains and stellar nucleosynthesis and for carefully reading the manuscript, which has been essential for corrections and improvements. I also thank Sachiko Amari, Claudio Ugalde, Amanda Karakas, Larry Nittler, Falk Herwig and Andy Davis for sharing comments, pictures and data tables, Richard Stancliffe for proofreading the manuscript, Pierre Lesaffre for help with plotting, and Mariateresa Chiesa for a careful check of the exercises. Especially, I thank Brett Hennig, who played the role of the general reader with much dedication throughout the preparation of the manuscript, and for constant support during the work.

I gratefully acknowledge the support during the time this book has been written of the Institute of Astronomy of the University of Cambridge, through the Particle Physics and Astronomy Research Council Theory Rolling Grant.

M. Lugaro

Contents

Preface   vii

1. Meteoritic Presolar Grains and Their Significance   1

    1.1 Presolar isotopic signatures and their carriers   3
    1.2 The discovery of presolar stellar grains   8
    1.3 Meteorites carrying stellar grains 10
    1.4 Types of presolar grains 11
         1.4.1 Diamonds 13
         1.4.2 Silicate grains 14
         1.4.3 Silicon carbide grains 14
         1.4.4 Graphite grains 16
         1.4.5 Oxide grains 17
         1.4.6 Silicon nitride grains 17
    1.5 New information from presolar grains 18
         1.5.1 Stellar evolution, nucleosynthesis and mixing 18
         1.5.2 Physical and chemical properties of the gas around
                stars and supernovæ 20
         1.5.3 The interstellar medium, molecular clouds and early
                solar system 22
    1.6 Outline 24
    1.7 Exercises 24

2. Basics of Stellar Nucleosynthesis   25

    2.1 Hydrogen burning, and the life of most stars 27
         2.1.1 The pp chain 29
         2.1.2 The CNO, NeNa and MgAl cycles 32

2.2 Helium burning, and the evolution of stars of low mass . . .   36
2.3 The α process: C, Ne and O burnings, and the evolution of
     stars of high mass   39
2.4 The e process: Si burning, and supernova explosions   40
2.5 The production of elements heavier than Fe   45
     2.5.1 The s process   49
     2.5.2 The r process   51
     2.5.3 The p process   56
2.6 Exercises   57

3. Laboratory Analysis of Presolar Grains   59

3.1 The isolation of diamond, graphite and SiC grains   59
3.2 Looking at presolar grains   62
3.3 Isotopic measurements with mass spectrometers   63
     3.3.1 Noble-gas extraction   65
     3.3.2 Secondary Ion Mass Spectrometry (SIMS)   65
     3.3.3 The advent of Resonant Ionization Mass
            Spectrometry (RIMS) in trace element analysis . . .   68
3.4 Location and analysis of rare types of presolar grains   71
3.5 Concluding remarks   74
3.6 Exercises   75

4. The Origin of Presolar SiC Grains   77

4.1 Classification of SiC grains on the basis of their C, N and Si
     compositions   77
4.2 Where did mainstream presolar SiC grains come from? . . .   81
     4.2.1 Theoretical modelling of AGB and C(N) stars   83
4.3 Carbon and nitrogen in mainstream SiC grains and in AGB
     stars   88
4.4 The Ne-E(H) anomalous component   94
4.5 The presence of ^{26}Al   98
4.6 The puzzle of the silicon isotopic composition of mainstream
     SiC grains   99
4.7 Titanium isotopic composition of mainstream
     SiC grains   105
4.8 A, B, X, Y and Z: The minor SiC grains populations   108
     4.8.1 The Y and Z populations   108
     4.8.2 The A and B populations   109

4.8.3 The X population 110
4.9 Exercises 111

5. Heavy Elements in Presolar SiC Grains   113

5.1 Modelling the *s* process in AGB stars 114
    5.1.1 The neutron source in AGB stars 115
    5.1.2 The production of a ^{13}C pocket 117
    5.1.3 The current model 121
    5.1.4 The neutron flux in the ^{13}C pocket 123
    5.1.5 The neutron flux in the thermal pulse 124
5.2 SiC grain data and the *s* process in AGB stars 126
    5.2.1 Class I: Isotopic ratios involving *p*-only and *r*-only
          isotopes 126
    5.2.2 Class II: Isotopic ratios involving isotopes in
          local equilibrium 130
    5.2.3 Class III: Isotopic ratios involving isotopes
          with magic neutron numbers 132
    5.2.4 Class IV: Isotopic ratios involving isotopes depending
          on branchings 137
    5.2.5 Class V: Isotopic ratios involving isotopes
          produced by radioactive decay 141
5.3 The heavy noble gases: Kr and Xe 142
5.4 Exercises 147

6. Diamond, Graphite and Oxide Grains   151

6.1 Diamond 151
6.2 Graphite 154
6.3 Oxide grains 157
6.4 Exercises 162

Appendix A   Glossary   165

Appendix B   Solutions to Exercises   173

B.1 Chapter 1 173
B.2 Chapter 2 173
B.3 Chapter 3 175
B.4 Chapter 4 177
B.5 Chapter 5 179

B.6 Chapter 6 181

Appendix C   Selected Books and Reviews for Quick Reference   183

C.1 Presolar grains 183
C.2 Stellar evolution and nucleosynthesis 183
C.3 AGB stellar evolution and nucleosynthesis 184

Bibliography   185

Index   207

Chapter 1

Meteoritic Presolar Grains and Their Significance

One of the most fascinating discoveries of the last fifty years is that much
of the material that constitutes our world and ourselves came from the
stars. All elements heavier than hydrogen and helium are produced in stellar interiors where temperatures of several million degrees allow nuclear
fusion reactions to occur (*nucleosynthesis*) [47, 50, 131, 287]. At the end
of the life of a star the newly formed elements are ejected into the interstellar medium from which new stars are born. Material in the Universe
is processed in a continuous cycle and the chemical composition of galax-
ies evolves (Fig. 1.1), so that different compositions are present within a
galaxy at different times and locations. The birth of the theories of stellar
nucleosynthesis and Galactic chemical evolution owed much to the availability of spectroscopic observations of the abundances of the elements in stars
[184, 185], and to the first compilation of the distribution of the abundances
of the elements in the solar system compiled in 1956 [267].

Until a few decades ago it was believed that the abundances of the
different isotopes of each element[1] in the solar system were completely
homogenised, because all the material present in the protosolar nebula –
the planetary disk formed during the gravitational collapse from which the
solar system originated – was very well mixed. The abundances of different
elements do vary in different locations of the solar system depending on their
chemical properties, such as their ability to form molecules and condense
into solids, and on the thermal conditions at which different solar-system
bodies were formed. For example, the amount of iron (Fe) condensed into
rocks on Earth is much higher than the amount of Fe that remained as a
gas in the atmosphere, because Fe can condense into solid material. The

---

[1]Nuclei belonging to the same element, i.e. characterised by a given number of protons,
but with different numbers of neutrons

1

New stars
form in the
interstellar
medium (ISM).

The material
in the stars is
processed by
nuclear reactions
and its composition
changes.

At the end of their
life the stars return
most of the material
to the ISM by stellar
winds or a supernova
explosion.

The composition of
the ISM is modified.
New stars form
and the same cycle
starts again.

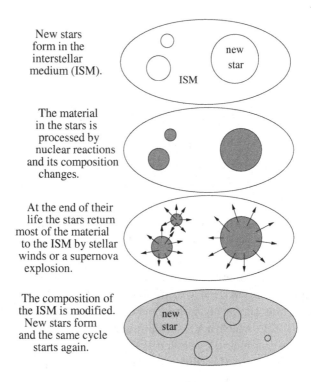

Fig. 1.1   Schematic representation of the recycling of material in a galaxy and the resulting evolution of the composition of stars and the interstellar medium.

amount of the noble gas Xe, instead, is much higher in the atmosphere since
this gas does not easily condense into solids.

However, if complete homogenisation of the solar material is assumed,
then the fraction of Fe, or any other element, made up by each of its stable
isotopes: ^{54}Fe, ^{56}Fe, ^{57}Fe, and ^{58}Fe is the same in every corner of the
solar system (with, for example, ^{56}Fe representing 92% of all Fe in the
solar system, and ^{58}Fe only accounting for 0.3% of it). This is because the
chemical reactions and physical processes that produced the material in the
solar system occurred at temperatures of at most a few thousand degrees,
and could have changed the isotopic composition of an element only at the
level of a few parts per thousand. Observed variations much larger than that
can only be present if the material was *originally* anomalous in its isotopic
compositions. Large isotopic anomalies, in fact, can only be produced by
nuclear reactions, occurring at temperatures of million degrees, by which

the structure of the atomic nucleus, i.e. the number of neutrons present in the nucleus, is modified.

Recently, the idea of a completely uniform isotopic composition in the solar system has been challenged by the discovery of small amounts of material that have "anomalous" or "exotic" isotopic compositions, i.e. differing from that commonly observed in the solar system. This material is mostly recovered from meteorites, relatively small extra-terrestrial rocks that fall onto the Earth. Since the observed large variations in the isotopic composition of such anomalous material cannot have been produced during the formation of the solar system, it is believed that this exotic material carries the signature of processes that predated the formation of the protosolar nebula, and it is hence labelled as "presolar".

In the next section, different types of presolar material are presented, and the rest of this chapter is focused on one particular type of presolar material: *stardust*, the meteoritic stellar grains that are the topic of this book. A brief history of their discovery is told in Sec. 1.2, the different types of stellar grains recovered so far from meteorites are introduced in Sec. 1.4. In the last section, 1.5, the many different types of information that can be derived from the study of presolar stellar grains are summarised.

1.1 Presolar isotopic signatures and their carriers

There are several types of presolar isotopic signatures, which are summarised in Table 1.1. The first type are represented by some meteoritic solids showing the consequences of the radioactive decay of unstable nuclei such as ^{26}Al, ^{41}Ca, ^{60}Fe and ^{107}Pd, which have relatively long half-lives[2], between 0.1 and 100 million years. The signature of the presence of radioactive nuclei in the early solar system shows up today in excesses of the abundance of their daughter nuclei, i.e. those they decay into. For example, large excesses in ^{26}Mg with respect to solar system material, but not in ^{24}Mg and ^{25}Mg, have been measured in small clumps of calcium- and aluminium-rich material, Ca- and Al-rich inclusions (CAIs, see Fig. 1.4) from the Allende meteorite [165]. These ^{26}Mg excesses are attributed to radioactive decay of the unstable nucleus ^{26}Al that must have been initially present in the solar system and incorporated into CAIs at the time of their formation.

---

[2]The time by which the abundance of a radioactive nucleus decreases by half of its initial abundance because of the decaying process.

Table 1.1   Types of presolar isotopic signatures.

anomaly	carrier	source
1) excesses of nuclei produced by radioactive decay	Ca- and Al-rich inclusions (CAIs) in primitive meteorites	presence of radioactive nuclei in the early solar system
2) small isotopic anomalies (order 10^{-4})	CAIs	small inhomogeneities in the solar nebula, or chemical effects
3) anomalies of deuterium and ^{15}N	primitive meteorites and Interplanetary Dust Particles (IDPs)	chemical processes in the molecular cloud where the Sun formed.
4) isotopic anomalies of many elements, up to four orders of magnitude	stellar grains recovered from primitive meteorites	stellar nucleosynthesis

Interesting questions are posed regarding where these radioactive nuclei have come from and their origin is still much debated. Initial abundances derived for radioactive nuclei with relatively long half-lives – higher than about 10 million years – are explained as the result of the evolution of the abundances in the interstellar region where the Sun was born due to equilibrium between Galactic production in stars and radioactive decay [192, 248]. Some light nuclei with shorter half-lives could have been generated during the active early phases of the formation of the Sun by some sort of interaction with accelerated particles (solar cosmic rays) [166], in which case the origin of these nuclei is not truly "presolar". Beryllium-10, in particular, could have been produced by this type of process, but occurring before the formation of the Sun, by interaction with Galactic cosmic rays [84]. Alternatively, some of these nuclei could have been produced by an event that occurred close to and just before the formation of the Sun and polluted the protosolar nebula with radioactive material, such as a supernova explosion [53, 192] or the winds of a red giant star [290]. If some of these nuclei came via stellar winds or explosion ejecta, then such an event could also be related to the triggering of the formation of the solar system. Interstellar shock waves could have initiated the collapse of the presolar gas cloud to form the Sun [43]. The study of these anomalies can ultimately shed light on the dynamical sequence of events that produced our Sun and Earth.

When material with anomalous composition is mixed and diluted with non-anomalous material, the sign of the early presence of *unstable* nuclei discussed above is not lost because it shows up in excesses of the abundances of the stable daughter nuclei they decay into. On the contrary, it is much more difficult to recover anomalies in the composition of *stable* nuclei, when anomalous material is mixed and heavily diluted with material having composition like that of the bulk of the solar system.

Small anomalies in the composition of stable nuclei of calcium, titanium, and iron are shown by CAIs [73] (presolar material of type 2 listed in Table 1.1). The first accepted sign of the presence of presolar material in meteorites was an observed $+4\%$ deviation from the abundance of solar ^{16}O in CAIs [72]. However, there is the possibility that this anomaly could have been produced in the solar system by chemical effects [275].

The third type of presolar material is identified by anomalous abundances of deuterium ($D=^2H$) and ^{15}N observed in primitive meteorites as well as in Interplanetary Dust Particles (IDPs), i.e. tiny meteorites (diameter $< 50 \mu m$) collected in the stratosphere. Observed D/H ratios are up to 10 times higher than in terrestrial rocks [189, 186]. These anomalies are within the range observed in some molecular clouds, which are cool and dense regions of the interstellar medium where atoms tend to combine into molecules. High D/H ratios are due to isotopic fractionation occurring during chemical reactions between ions and molecules at very low temperature, and to the fact that the relative mass difference between D and H is very high [277]. Also ^{15}N excesses of up to 50% of the terrestrial standard are observed in IDPs, which can also associated with chemical fractionation, if reactions take place at very low temperatures.

The fourth type of presolar material is that which is discussed in the rest of this book: *stardust*, also called stellar, or presolar, dust grains. These grains have isotopic compositions very different to those commonly measured in the solar system: they display enormous anomalies of up to four orders of magnitude in their isotopic compositions. These are too large to be attributed to chemical or physical fractionation and could only have been produced by nuclear reactions, which occur in stars. After being ejected into the interstellar medium presolar stellar grains were incorporated in the protosolar nebula and then survived the formation of the solar system without being destroyed. These grains existed in their current form before the solar system was born and have kept their own individuality and composition until today. General reviews on presolar stellar grains can be found in Refs. [21, 68, 171, 301, 302], while a large number of detailed reviews on a variety of related subjects are collected in Ref. [28].

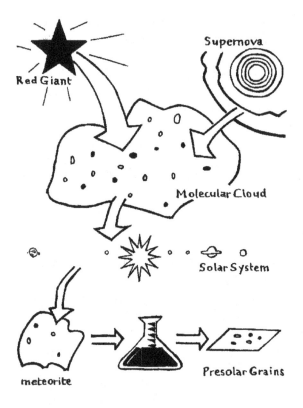

Fig. 1.2   Schematic representation of the journey of presolar grains from their site of formation around stars to the laboratory (courtesy Larry Nittler).

A schematic representation of the journey of stellar presolar grains from their circumstellar site of formation to the laboratory is shown in Fig. 1.2. It begins with their birth, as dust grains forming in the gas surrounding stars. For example, favourable locations for the formation of dust are the wind-driven extended envelopes of red giant stars, and apparently the majority of known presolar grains came from these stars. The outer regions of giant stars, or circumstellar shells, are cool enough ($\simeq$ 1500 K) for the gas to condense into grains. The presence of heated dust in circumstellar shells, which is observationally associated with strong mass loss from the star, was known and studied well before the discovery of stellar grains in meteorites, as it causes the infrared excess seen in the spectrum of many giant stars: the dust makes the circumstellar shells opaque so that the light coming from the star is absorbed and re-emitted in the infrared. Other sites where dust can form are in the cooled ejecta of nova and supernova explosions. The stellar dust, together with most of the gas that formed the star, is ejected into the interstellar medium by the stellar winds or the nova or supernova explosion. About 1% of the mass of the interstellar medium is made up of this *cosmic dust*, which causes interstellar extinction, i.e. the dimming of light from stars. When the Sun formed in a molecular cloud, some of the grains present in the protosolar nebula were trapped in asteroids. Fragments resulting from the impact of smaller rocks on an asteroid can have their course deflected in such a way that they reach the surface of the Earth. These extra-terrestrial pieces of rock are the meteorites from which presolar grains are recovered today[3]. Since presolar grains carry the signature of the composition of the gas that surrounds stars, the laboratory analysis of their composition and structure provides invaluable information on the stars, the astrophysical site of their origin.

The discovery of stellar grains in meteorites has created a new field of astronomy, where scientists from different disciplines, from nuclear physics to astronomy and chemistry, are required to work together. In this new field information and constraints on theories of Galactic evolution and stellar nucleosynthesis come from measurements on dust grains performed in the laboratory.

While the bulk of the solar material came from many different stars because of Galactic chemical evolution, each presolar grain carries the signature of its site of formation around its parent star. Hence the composition of stellar grains gives us a unique opportunity to study the composition of a single star, rather than a mixture of them. Thus, data from presolar grains are similar to the spectroscopic observations of stellar atmospheres, however, they are conceptually and practically different in several ways. While stellar observations usually deal with elemental abundances, and only in rare cases yield the isotopic composition of the star, the laboratory analysis of presolar material yield data on isotopic compositions. Theories have thus to be tested against information about isotopic compositions, which presents more detailed constraints than elemental data. Moreover, the laboratory measurements of presolar material, with error bars in some cases as low as a few percent, are typically much more precise than spectroscopic observations, with typical error bars of a factor of two.

On the other hand, while we do know from which star spectroscopic observations are derived, and hence we automatically have information such

---

[3] Other types of meteorites came from the Moon and Mars.

as its spectral type and luminosity, that allows us to classify the star in our
theoretical framework, we do not know *a priori* in which astrophysical site
the grain was produced. The first challenge of measured presolar compositions is to understand which conditions made them possible. Only after
this is reasonably well established can the data be used to constrain the
theories. Stellar observations and the analysis of presolar material complement each other in bringing new challenges to our knowledge of the stars
and our Galaxy.

1.2   The discovery of presolar stellar grains

The first hint of the existence of presolar stellar grains appeared in the
1960s from the analysis of the composition of the noble gases neon (Ne)
and xenon (Xe) in primitive meteorites. In spite of the fact that noble
gases have very low concentrations in other materials, hence they are also
known as *rare* gases[4], they are found as trapped or implanted bubbles inside
materials and can be extracted during heating experiments. Analysis of the
composition of Xe [233] and Ne [33] in old carbonaceous meteorites showed
the presence of "exotic" components with isotopic compositions completely
different from the bulk of solar system material.

Two exotic components were found for Xe: Xe-HL, which stands for Xe
heavy and *light* [233], and Xe-S, which stands for Xe *s process* [260] (see
Fig. 1.3). In the case of Xe-HL, the light isotopes ^{124}Xe and ^{126}Xe and
the heavy isotopes ^{134}Xe and ^{136}Xe are especially enhanced with respect
to the solar composition. The production of these isotopes is related to
two nucleosynthesis processes: the *proton*-capture process (*p* process, see
Sec. 2.5.3) for ^{124}Xe and ^{126}Xe, and the *rapid* neutron-capture process
(*r* process, see Sec. 2.5.2) for ^{134}Xe and ^{136}Xe. In the case of Xe-S,
instead, ^{128}Xe and ^{130}Xe are the most enhanced isotopes. Their production
is related to the *slow* neutron-capture process (*s* process, see Sec. 2.5.1).

Anomalous Ne was found to be very much enriched in ^{22}Ne with respect
to solar material with ^{22}Ne/^{20}Ne and ^{22}Ne/^{21}Ne ratios up to a thousand
times higher than in the Sun. This component was named Ne-E (as the
letters A, B, C and D were already taken for other Ne components), and
further distinguished in two types: Ne-E(L) and Ne-E(H) because of differ-

---

[4]Noble gases He, Ne, Ar, Kr and Xe, are the most volatile elements, they do not form any
chemical compounds and condense only at extremely low temperatures. This is because
their atomic structure is very stable and hence they do not react easily with other atoms.

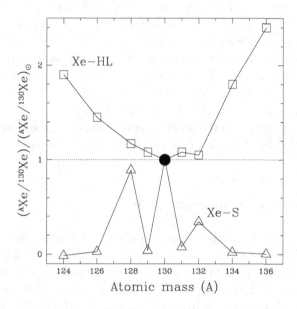

Fig. 1.3 Isotopic composition of the two Xe exotic components present in meteorites. All isotopic ratios are calculated with respect to ^{130}Xe and divided by the corresponding ratios in the solar system, so that data for ^{130}Xe always equals 1 (full circle). In the case of Xe-HL (open squares) all isotopes are more abundant than ^{130}Xe, with respect to solar, particularly at the two endpoints of the isotopic range. In contrast, in the case of Xe-S (open triangles) ^{130}Xe is the most abundant isotope, followed by ^{128}Xe.

ences in the heating temperature (*lower* or *higher* than 1200 °C) at which they are released. Also, the density of the type of grains that host them turned out to be lower or higher than 2.3 g/cm^3 for the two components, respectively.

These anomalous compositions were not plausibly explained as a product of processes that occurred in the solar system, but seemed rather to be the product of nuclear reactions in stellar interiors. This led to the quest of the isolation of these components, so to identify their carriers inside the meteorites.

It took approximately ten years for Anders and his colleagues at the University of Chicago to identify and extract the minerals within the meteorites responsible for carrying the exotic noble gas components [12, 270]. In the case of Xe-S and Ne-E(H) the carrier was found to be silicon carbide (SiC) grains [30, 306] and in the case of Xe-HL the carrier was found to be diamond [169] (see also Ref. [270]). In the case of Ne-E(L) the carrier was found to be graphite [6]. The final isolation process has been described by Anders as "burning down the haystack to find the needle", as it involves dissolving the meteoritic rock using appropriate harsh acids until only the presolar dust remains (as will be described in Sec. 3.1). Fortunately, many presolar grains are made of hard and resistant material, able to survive all these destructive processes. On the other hand, they are microscopic and their abundance relative to the other minerals composing the meteorites is very small, typically of the orders of parts per million (by mass).

1.3   Meteorites carrying stellar grains

Presolar stellar grains have been found in all classes of primitive *chondrite* meteorites. Chondrite meteorites represent more than 80% of all meteorite falls and are characterised by the presence of *chondrules*, small spheres of average diameter $\simeq$ 1 mm of previously melted minerals that have come together with other mineral matter to form a solid rock (see Fig. 1.4).

The abundance of presolar grains in a chondrite meteorite correlates with the percentage of the rock that is composed by the *matrix*, the amalgam of amorphous material and crystals of very small dimensions, of the order of 10^{-6} m, not visible with an optical microscope. This implies that presolar grains are situated in the matrix, representing some of the microcrystals of which the amalgam is formed. The types of chondrite meteorites that are mostly composed of matrix and hence contain the largest abundance of presolar grains, are the subgroups CI and CM2 belonging to the rare class of *carbonaceous* chondrites, of which examples are Orgueil (CI, France 1964) and Murchison (CM2, Australia 1969). These meteorites contain carbon, of which some is in the form of organic compounds, and are believed to be the oldest stones formed in the solar system. Their bulk composition is almost identical with that of the Sun, thus they represent a most reliable source from which to derive standard solar abundances of elements and isotopes (see e.g. [20]).

The abundance of presolar grains in different chondrites normalised to their matrix content depends on the grade of *metamorphism* of the meteorite, i.e. how much the meteorite has endured processes characterised by variations in temperature, pressure and the presence of fluids, which can cause a variation of the structure and composition of the rock. Meteorites of types CI and CM2 are also the least metamorphosed type of chondrites hence they contain the largest absolute number of presolar grains.

Fig. 1.4 Photograph of a stone from the Allende meteorite, which fell in Mexico in 1969. The small squares in the background grid have 1 mm size (courtesy Eric Twelker). Allende is a carbonaceous chondrite of class CV3. Chondrules with size of the order of mm (10^{-3} m) stand out against the darker matrix, where presolar grains of the size of μm (10^{-6} m) are located. The clumps of white material are calcium-aluminium-rich inclusions (CAIs) where the signature of the early presence of the radioactive nucleus ^{26}Al has been discovered (Sec. 1.1).

1.4 Types of presolar grains

In Table 1.2 is presented a list of the types of presolar grains recovered so far and their abundances in the Murchison meteorite, in units of parts per million (ppm). The most abundant type is diamond, other relatively abundant carbon-bearing minerals are silicon carbide and graphite. Silicate grains are probably the second most abundant presolar grains, though data are not yet available for the Murchison meteorite since these grains have only very recently (in 2004) been discovered in the Acfer 094 and North West Africa 530 meteorites. In some meteorites, oxide grains such as corundum are more abundant than silicon carbide, thus the order in which the grains are listed in Table 1.2, which is only based on available analysis of the Murchison meteorite, should be considered as indicative. The question of which type of presolar grains are in general more abundant will only be settled through further analysis.

Graphite and SiC grains also contain tiny subgrains of Ti, Zr, Mo, Ru and Fe-carbides [29] as well as subgrains of Fe-Ni metals [75]. Also polycyclic aromatic hydrocarbon (PAH) molecules have been found in many graphite grains [187].

Presolar grains are all refractory, which means that, at high temperatures, roughly between 1300 and 2000 K, they can condense directly from the gas phase. The condensation sequence of minerals depends on the initial composition of the gas, and indeed mainly on the C/O ratio. If C/O < 1, all the carbon is locked up in carbon monoxide (CO) molecules, which have a very strong bond and are stable at high temperature, and the condensed minerals are mostly oxides and silicates. If C/O > 1, instead, all the oxygen is locked up in CO molecules and carbon compounds can condense, such as graphite and carbides. Since in the solar system C/O ≃ 0.4, carbon bearing minerals could not have condensed in the protosolar nebula. Hence, these types of meteoritic grains are virtually all of presolar origin.

Table 1.2 Types of presolar grains, abundance in the Murchison meteorite [132], and typical size.

type	abundance (ppm)	size (μm)
diamond	>750[a]	0.002
silicates	—[b]	0.1 – 0.5
silicon carbide (SiC)	9[a]	0.1 – 20
spinel ($MgAl_2O_4$)	1[c]	0.5
graphite	>2[d]	0.8 – 28
corundum (Al_2O_3)	∼0.005[e]	0.5 – 4
silicon nitride (Si_3N_4)	>0.002[f]	∼ 1

[a] Found and measured in ≃ 40 chondritic meteorites. Abundances vary with the matrix content and metamorphic degree of the meteorite.
[b] Identified in the Acfer 094 and North West Africa 530 meteorites with contrasting abundance estimations of ≃ 25 ppm [200], 30 ppm [199] and 75 ppm [194].
[c] Also identified in other chondritic meteorites such as Murray, Orgueil and Acfer 094.
[d] Also identified in nine other chondritic meteorites.
[e] Also identified in four other chondritic meteorites, with abundance up to 0.2 ppm.
[f] Inaccurately known, also identified in four other chondritic meteorites.

1.4.1 Diamonds

The most abundant presolar grains are very small diamonds of the size of nanometers (10^{-9} m), hence they are called *nanodiamonds*. These grains are far more abundant than any other presolar grain. For example they constitute almost 6% of the total carbon in the Murchison meteorite. Presolar diamonds carry the exotic Xe-HL component related to p- and r-process nucleosynthesis [270]. Since these processes are predicted to occur in massive stars exploding as supernovæ, presolar diamonds probably have a supernova origin. The implications of this origin will be discussed in Sec. 6.1. The presence of the Xe-HL component, and their Te and Pd compositions, are the only information from the nanodiamonds that points to their presolar origin. However, this origin can be strictly applied only to a small fraction of the diamond grains because the concentration of Xe is extremely low so that only about one nanodiamond in each million contains an atom of anomalous Xe. Moreover, because diamond nanocrystals are too small to be analysed one by one, their carbon isotopic composition can be measured only in bulk, i.e. in collections of large numbers of grains. In this way it is only possible to obtain data on the carbon composition averaged on millions of grains, which happens to be very close to solar [239]. Of course, this does not necessarily mean that all the grains have solar carbon composition, because extreme compositions would cancel each other out in the averaging process. The $^{14}N/^{15}N$ ratio is on average about 35% higher than the terrestrial value, but in agreement with the ratio observed in Jupiter [221] . As for noble gases in meteoritic diamonds, a detailed study of the different components and the implications for their origin can be found in Ref. [134].

The favoured mechanism for the formation of nanodiamonds has been identified in a chemical vapour-deposition-like process occurring at low pressure [78], by which material in a vapour state condenses through chemical reactions, rather than a high-pressure shock-induced metamorphism that produced, for example, diamonds in meteorite craters. This is consistent with condensation in cool stellar atmospheres. For the inclusion of noble gases in the diamonds, ion implantation is the most likely mechanism, which would also be consistent with the fact that the concentration of noble gases increases with the grain size [285].

In summary, it is not known which fraction of the diamond grains are actually presolar. Some of them could have formed in the inner regions of the solar system [76]. This is suggested by the fact that nanodiamonds are mostly absent in Interplanetary Dust Particles of cometary origin, and comets are thought to have formed further out in the early solar system and to be older than asteroid parent bodies. Moreover diamonds are detected within the accretion discs of young stars.

1.4.2 *Silicate grains*

Before 2004, the inability to find presolar silicate grains in meteorites was puzzling. This was because the major oxide phases observed around red giant stars are silicates, represented by SiO, other Si-based minerals such as olivine and pyroxene, and amorphous silicate grains [80, 291]. However, no presolar silicates were to be found in meteorites. Problems are that silicates are more likely to be destroyed by chemical processing during the life of the meteorite, and that presolar silicate are very difficult to locate among the abundant silicate of solar origin that constitute the main part of meteorites. Moreover, silicates were destroyed by most of the chemical treatments used to prepare meteoritic residues. Presolar silicates were also difficult to detect because of their small size.

Thanks to the advent of the NanoSIMS instrument (Sec. 3.3.2) it is now possible to identify and analyse presolar grains of smaller sizes than was possible before. The existence of presolar silicate grains in the solar system has thus been confirmed by their discovery within Interplanetary Dust Particles [188], where they are quite abundant: $\simeq 5500$ ppm in cluster IDPs, and also in the carbonaceous chondrites Acfer 094 and North West Africa 530 [194, 199, 200]. The estimated abundance of silicates in these meteorites varies from $\simeq 25$ to $\simeq 75$ ppm, exceeding that of any other presolar material, except diamond. It is foreseen that by using improved techniques many more of these grains will be collected from various sources in the near future.

1.4.3 *Silicon carbide grains*

Silicon carbide grains (SiC) are large enough to allow the analysis of single grains, since their size varies from a fraction to a few tens of μm. They are also relatively easy to extract from meteorites, with respect to the other types of presolar grains, and thus several thousands of them have been analysed to date. SiC grains typically have shapes bounded by crystal planes, with more or less pitted surfaces, likely due to the harsh treatments

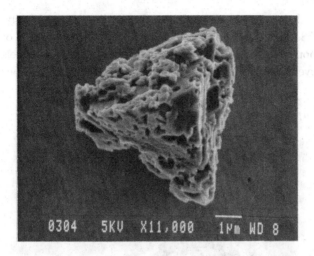

Fig. 1.5 High-resolution scanning electron microscope image of a presolar SiC grain of size $\simeq$ 6 μm from the Murchison meteorite. The $^{12}C/^{13}C$ ratio of this grain is 55, while in the solar system is 89 (courtesy Sachiko Amari).

to which they were exposed during the extraction (Fig. 1.5[5]). Even if SiC can crystallise in many different ways, 80% of presolar SiC grains have cubic form (β-SiC, as opposed to α-SiC, which refers to a variety of different structures) and the remaining 20% have hexagonal form [77].

Since their discovery, presolar SiC grains have been extensively studied (see e.g. [122, 127]) and a relatively large amount of information is available, which will be discussed in detail in Chapters 4 and 5. Based on their C and Si composition, SiC grains have been classified into several populations, the largest of which (*mainstream* SiC) comprises more than 90% of the grains. The remaining SiC grains are classified in other five small populations: A, B, X, Y and Z (Sec. 4.8). Presolar SiC contains impurities dominated by N, Al and Ti, and trace elements with low concentrations such as Mg, Ca, Zr, Mo, Ru, Ba, Nd.

These grains are the carriers of the Xe-S exotic component, which is produced by s-process nucleosynthesis. Enhancements in the elements produced by the s-process are observed in red giant stars on the Asymptotic Giant Branch (AGB), and therefore most SiC grains are believed to have formed in the carbon-rich envelopes of AGB stars. In fact, the emission line

---

[5]Reprinted from Chemie der Erde, Lodders & Amari, Presolar grains from meteorites: Remnants from the early times of the solar system, to appear, Copyright(2005), with permission from Elsevier.

at 11 μm – characteristic of β-SiC – is observed spectroscopically in these stars [74, 258, 259, 282]. The fact that SiC grains are the carriers of the Ne-E(H) component also fits into this scenario as theoretical models predict that the envelopes of AGB stars are also enriched in ^{22}Ne (see Sec. 4.4).

1.4.4 Graphite grains

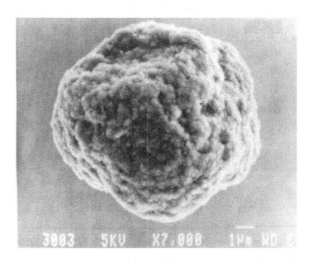

Fig. 1.6 High-resolution scanning electron microscope image of a presolar graphite grain of cauliflower-like morphology and size $\simeq$ 6 μm from the Murchison meteorite. Presolar graphite grains always show spherical appearance (courtesy Sachiko Amari).

Like SiC grains, graphite grains are large enough to be analysed singularly. However, their extraction procedure is more complex than that of other grains, because graphite has chemical and physical properties similar to those of other carbonaceous compounds present in the meteorite. Moreover, trace elements are present in extremely low abundances, which makes their analysis challenging. A few hundreds graphite grains have been analysed to date and have been classified according to their morphologies and densities. Round grains, which comprise more than 90% of graphite grains, have isotopically anomalous carbon and hence are clearly of stellar origin. Their density is in the range 1.6–2.2 g/cm^3 and they have two different external appearances: cauliflower-like, consisting of aggregates of smaller grains (Fig. 1.6) and more abundant among grains of low density, and onion-like, consisting of concentric layers of graphitised carbon and more abundant among grains of high density. About one third of all presolar graphite grains have low densities (1.6–2.05 g/cm^3) and appear to have originated from supernova explosions [123, 281]. Higher-density grains could have originated from a range of stellar environments. Details on this type of grains and their origin are presented in Sec. 6.2.

1.4.5 *Oxide grains*

Until 2003, corundum (sapphire and ruby, Al_2O_3) was believed to be the most abundant presolar oxide grain. However, with the NanoSIMS instrument (Sec. 3.3.2) it is now possible to identify and analyse presolar grains of smaller sizes, and it has been recently found out that spinel ($MgAl_2O_4$) is the most abundant presolar oxide grain [304]. This result was missed before because spinel grains have average sizes smaller than corundum grains. A few hibonite grains (with composition $CaAl_{12}O_{19}$) and one titanium oxide grain (TiO_2) of presolar origin have also been recovered [58, 59, 212]. Oxide grains are resistant to the chemical treatments used to isolate the carbonaceous grains and they are present in meteoritic residues together with SiC grains. However, presolar oxide grains are more difficult to locate because the majority of oxide grains in meteorites formed in the solar system, where $C/O < 1$. Only a small fraction of oxide grains are of presolar origin and special techniques are needed to recognise them (see Sec. 3.4). These types of grains have not been traced through the presence of noble gases, but have been recognised during analysis of acid-resistant meteoritic residues because of their anomalous composition.

Oxide grains have been separated into distinct groups, based on their oxygen and aluminium isotopic ratios. The $^{26}Al/^{27}Al$ is derived for the time when the grain formed by estimating the initial presence of ^{26}Al from the radiogenic abundance of ^{26}Mg. The composition of most of these grains suggests that they have formed around red giant and AGB stars, as will be discussed in Sec. 6.3.

1.4.6 *Silicon nitride grains*

Also silicon nitride grains (Si_3N_4) have been identified during analysis of meteoritic residues because of their anomalous composition [215]. The condensation of silicon nitride requires $C/O > 1$ and a high nitrogen concentration. The composition of these type of grains is very similar to that of SiC grains belonging to the X population (Sec. 4.8.3) and points to a supernova origin of Si_3N_4 grains.

1.5   New information from presolar grains

When considering the different astronomical sites through which presolar grains journey, as represented in Fig. 1.2, it is clear why these grains represent not only a new field of astronomy, where dust from stars is analysed in the laboratory, but also a new scientific field requiring the common effort of scientists from very different disciplines. These disciplines range from nuclear physics, to theoretical astrophysics, observational astronomy, cosmochemistry and the laboratory analysis of materials. The information that we can extract from presolar grains is summarised in Table 1.3 and discussed in the rest of this section.

Table 1.3   Information from presolar grains relating to their site of formation and their journey through space.

site	information
Circumstellar regions	- initial composition of the star (Galactic Chemical Evolution)
	- stellar thermal structure, nucleosynthesis and nuclear reaction rates
	- mixing processes inside red giant stars and during nova and supernova explosions
	- physical and chemical properties of the gas around stars
Interstellar medium	- destruction processes of cosmic dust
	- exposure to Galactic cosmic rays
Molecular cloud	- cloud and grain chemistry
Solar system	- survival of presolar material in the early solar system
Meteorite	- metamorphism processes of meteorites

1.5.1   *Stellar evolution, nucleosynthesis and mixing*

The very precise analysis of the isotopic composition of presolar grains represents a breakthrough in the field of stellar evolution and nucleosynthesis. During the chemical process of formation of molecules and grains around stars, isotopic fractionation effects, i.e. preferences in incorporating different isotopes of a given element, could have only produced very small isotopic anomalies, at a level of a few parts per thousand. Thus, the enormous range of variation in the isotopic compositions observed in presolar grains is an extremely precise record of the isotopic composition of their site of formation and must be explained by models of nucleosynthesis and mixing in stars.

The composition of the parent star of a grain is determined both by the initial composition of the star and the nucleosynthesis occurring inside the star itself. The initial composition is a complex function of the age of the star and the place where the star was born and can be predicted by Galactic chemical evolution calculations. These calculations model the continuous recycling of matter represented in Fig. 1.1, from the interstellar medium into new stars in which the matter is processed and transformed by nuclear reactions, and from the stars back into the interstellar medium. The aim of such studies is to follow the evolution of the composition of matter in various regions of a galaxy and in many generations of stars. Predictions are usually compared with observations of the composition of stars of different ages, and of the interstellar medium. Because some elements are not modified by nucleosynthesis in the parent stars of presolar grains, their composition in presolar grains is believed to record the initial composition of the parent stars, thus providing detailed constraints on models of the evolution of the Galaxy (see e.g. Sec. 4.6).

The stellar composition is further modified by the nucleosynthesis occurring inside the star itself. These modifications depend on the initial mass, composition and evolutionary phase of the star (see Chapter 2). Modifications are due to nuclear reactions and usually take place in the hot stellar interiors. They depend crucially on the thermodynamic structure of the star and the efficiency of nuclear reactions. Temperatures and densities of the different region of the star are computed using theoretical models of stellar evolution, while nuclear reaction rates are measured in the laboratory, or calculated theoretically. The composition of many elements in presolar grains show large variations due to the nucleosynthesis that occurred in their parent stars. The analysis of these effects provides constraints on the thermal structure of the stars and on nuclear reaction rates.

In order for the nucleosynthesis occurring in the hot deep layers of the star to be relevant to the composition of the dust grains forming in the much cooler outer regions, some mixing mechanism must be at work so that the processed material is carried from those deep layers to the surface of the star. In red giant stars this mechanism is referred to as *dredge-up* (see Secs. 4.2.1 and 4.3) because the envelope of the star, where most of the stellar mass is located and where the transport of energy occurs by convection (fluid circulation), periodically extends to deeper regions of the star and brings material to the surface. Other mechanisms by which processed material is mixed to the surface of red giant stars are known as *hot bottom burning* (see Sec. 4.2.1) and *extra mixing* or *cool bottom processing* (see Sec. 4.3). In this cases the mixing occurs continuously because the nuclear processes occur at, or just below, the base of the convective envelope.

Mixing processes in stars are related to the general astrophysical problem of the treatment of turbulent convection in stellar interiors, and can also be connected to structural asymmetries, stellar rotation and the presence of magnetic fields. A current limitation of the theoretical models of red giant stars is that they are performed in one dimension. This allows us to make the computation feasible and obtain a good description of the basic properties of the star, but does not represent a realistic way of modelling mixing phenomena. The composition of presolar grains give us constraints that can help us to better understand such complex mixing processes.

In supernovæ the mixing of material from inner to outer layers can be achieved prior to or during the explosion. For example, the trend shown by observations of the variation with time of the light from the very bright supernova SN1987A can only be explained if, because of hydrodynamical instabilities (turbulence), "fingers" of material from the inner regions, in particular the radioactive nucleus ^{56}Ni, were shot to the outer regions [108, 294]. Microscopic mixing of different supernova regions is required to explain the composition of presolar graphite and SiC-X grains, which show the signature of the nucleosynthesis occurring in supernovæ of type II [281], i.e. they are produced by the explosion of massive stars (see Sec. 2.4). In these grains are found nucleosynthesis products both from the inner supernova regions, such as ^{44}Ti and ^{28}Si, together with those from the outer regions, such as ^{15}N and ^{26}Al (Secs. 4.8.3 and 6.2). The mixing issue is also strongly connected to the mechanism by which the grains condensed. Two- and three-dimensional hydrodynamical calculations involving a large computational effort are needed to attempt to produce realistic models of the pre-supernova and supernova stages [24]. The existence and the composition of presolar grains from supernovæ can be used as a road map in the difficult task of understanding mixing phenomena in supernovæ.

1.5.2 *Physical and chemical properties of the gas around stars and supernovæ*

From the structural features and the elemental composition of presolar grains it is possible to obtain constraints on the properties of the circumstellar gas in which they formed. For example, smaller dust particles of Ti, Zr, and Mo carbides are found inside larger graphite and SiC grains. These crystals can form within initially homogeneous SiC grains, but not within graphite grains. They must have thus condensed before graphite. On the other hand, the fact that no SiC particles have been found trapped in graphite grains indicates that SiC grains must have formed after graphite [29]. The elemental abundances of various trace elements in SiC grains give information on the chemical composition of the gas around stars and its thermal properties. Refractory elements, which condense from gas into solids at high temperature, such as Zr and Nd, are believed to have condensed together with SiC, while volatile elements, such as noble gases, which condense from gas into solids at low temperature, are believed to have been ionised, accelerated and implanted into the grains. Implantation models have provided independent constraints on the velocities of the material around red giant stars and planetary nebulae [286], as will be briefly discussed in Sec. 5.3.

Condensation models assuming thermochemical equilibrium around red giant stars, combined with simple grain growth models, have shown that it is possible to achieve the required condensation sequence in a carbon-rich gas (from Ti, Zr, and Mo carbides to graphite and SiC) and have set limits on the pressure of the gas, which should be in the range of 0.1 to 100 dyne/cm^2, and on the C/O ratio, which should be in the range of 1.05 to 1.2, to obtain the observed grain sizes [172, 252]. These models have also been successful in explaining the observed condensation patterns of refractory trace heavy elements in SiC grains [173]. However, the densities required to form TiC before graphite and to produce grains of the observed size are much higher than those predicted to characterise the dust-forming regions in red giant stellar atmospheres. This might mean that around these stars there are regions of a higher density than predicted. Another possibility is that large grains preferably form in long-living disks around interacting binary systems rather than around single stars [150, 151].

Moreover, in stellar outflows thermochemical equilibrium is a simple approximation because of the dynamic role played by the expanding and cooling matter, and the heating effect of possible shock waves. Several studies consider red giant envelopes as places of pronounced non-equilibrium, both in their thermodynamic and chemical features (e.g. [57, 222, 250]). These models attempt to include all the processes involved in the production of the grains, from the formation of the molecules to the growth of the dust itself. The theory is complicated by the fact that there is not yet a clear and satisfactory description of the observed mass loss through stellar winds shown by red giant stars, which, together with the treatment of mixing, represents a major uncertainty in stellar evolution models of red giant stars. The observed mass loss is up to ten orders of magnitude higher than that of the Sun: from 10^{-8} solar mass $(M_\odot)$ per year for stars on the first red giant branch (see Sec. 4.2.1), to 10^{-7} $M_\odot/\mathrm{yr}$ during the Asymptotic Giant Branch phase, and up to 10^{-4} $M_\odot/\mathrm{yr}$ in the final phases of the life of the star. Dust formation certainly plays a role as it is believed that one of the mechanisms responsible for the mass loss is radiation pressure on dust grains, which then drag the gas along with them.

Supernovæ are also believed to be one of the major producers of dust in our Galaxy, but many problems are still open on how the dust condensation physically occurs. The existence of graphite and SiC-X grains with the nucleosynthetic signature of supernova nucleosynthesis is a challenge to the theoretical modelling of mixing and dust formation during supernova explosions. From the observational point of view, dust has been observed in the cooled ejecta of the supernova SN1987A [44, 294] and in supernova remnants [87]. The study of dust formation around supernovæ, in particular in relation to presolar grains from meteorites, has barely begun. It is a complex task since the condensation of dust cannot in principle be decoupled from the mixing phenomena occurring in supernovæ. Both processes working together are likely to have determined the composition observed in the grains. The condition $C/O > 1$ is possibly not strictly necessary to form carbon-rich minerals in a supernova environment because CO molecules can be disrupted by energetic electrons produced by radioactivity in the supernova [67, 69]. A recent analysis of the impact of supernova shocks on the formation and composition of SiC-X from supernovæ can be found in [81], together with a detailed analysis of the issues involved, as well as the many unsolved problems.

Dust formation has also been associated with nova outbursts [263]. A few SiC and graphite grains have been recovered showing the signature of nova nucleosynthesis [7] thus giving further proof that dust production occurs also around novæ.

1.5.3   *The interstellar medium, molecular clouds and early solar system*

Before the discovery of presolar grains the study of cosmic dust was based only on its effects on starlight. The detailed laboratory observation of the structure and the appearance of presolar grains, as well as the relative abundance of the different types of presolar grains complement astronomical observations in the study of the interstellar dust, its propagation and its survival in the interstellar medium [147]. During their time in the interstellar medium dust grains are subjected to destructive processes such as sputtering by shocks and stellar winds, and are also likely to interact with Galactic cosmic rays, which are ions, mostly protons, that travel across the Galaxy and are probably accelerated by supernova shocks. This interaction can result in nuclear spallation reactions, i.e. the detachment of nucleons or small nuclei from larger nuclei as a result of the impact of energetic particles, which would thus modify the composition of the grains. While such modifications are yet to be observed unambiguously, they could reveal information on the age of presolar grains since their effect would depend on how long the grains have resided in the interstellar medium (see Sec. 4.4).

The survival of presolar grains during the formation of the solar system, and inside the solar-system bodies where they have been trapped, also represents a precious piece of information for the understanding of the formation of the solar system. The different types of presolar grains could have been destroyed to various degrees by potentially destructive processes occurring during the formation of the Sun (see e.g. [183]). In the early phases the grains were exposed to the thermal radiation of the collapsing cloud. Subsequently, the grains suffered an accretion shock produced by the dissipation of the energy of the material falling from the original cloud onto the nebula [54].

Since presolar grains have been found in all primitive meteorites studied they must have been initially present in the protosolar nebula all through the region of formation of meteorites. However, variations in the relative abundance of different types of grains in different classes of chondrites are found. For example, the abundance of SiC in enstatite ($Mg_2Si_2O_6$) chondrites is higher that that of diamond, with respect to the Orgueil and Murchison meteorites, while the abundance of diamond is higher that that of SiC in chondrites of type CV3. These data are of great importance for probing the properties of the protosolar nebula in which meteorite parent bodies formed, because the relative abundance of different types of grains depends on the temperature and composition of the surrounding environment [132]. More experimental data are necessary both on the abundances of the grains as well as on the destruction criteria of different types of grains in different conditions.

1.6   Outline

In Chapter 2 some basics of stellar nucleosynthesis are given, which will serve as a tool for the understanding of the chapters to follow. In Chapter 3 the various laboratory techniques currently used to extract and analyse presolar grains are described.

Constraints derived from presolar grain data and applied to theoretical nucleosynthetic models of the composition of their parent stars, are discussed in detail in Chapters 4, 5 and 6. In particular, Chapters 4 and 5 focus on the information from presolar SiC grains, which are the best-studied type of presolar grains and have been widely analysed. In Chapter 6 the constraints arising from the other types of presolar grains, diamond, graphite and oxide grains, are presented.

1.7   Exercises

(1) How many atoms are present in two presolar grains of radius 0.01 μm and 1 μm, respectively?

(2) How many milligrams of diamond, SiC and graphite grains are present in one gram of the Murchison meteorite?

(3) How many atoms of anomalous Xe are present among ten billion atoms constituting presolar diamonds?

Chapter 2

Basics of Stellar Nucleosynthesis

In order to understand how the anomalous composition of presolar stellar grains was produced, first we have to learn what type of nucleosynthesis occurs in stellar conditions and how this can affect the composition of the regions where the grains formed.

The idea that elements could be formed in stellar environments was first put forward by Hoyle in 1946 [131]. The other main theories for the origin of the elements considered that the elements were built all together in a primordial prestellar phase of the Universe (see e.g. review by [5]). This idea was based on the assumption, unchallenged at the time, that the stars and the interstellar matter had a uniform and unchanging chemical composition. In the 1950s, however, more detailed observations of the composition of stars became available showing element abundances that differed from those in the Sun [105]. In particular, in 1952 Merrill was able to observe the presence of the heavy element technetium (Tc, $Z=43$[1]) in some types of red giant stars. Technetium is a heavy element with no stable isotopes and its longest-living isotope has a half-life of four million years. Since the time it takes for a star to evolve into a red giant is longer than that, any Tc initially present would have decayed. Its presence, hence, represented a clear sign that its production was occurring somehow within the star. This discovery urged theorists to develop new theories for the production of the elements in stars (for Tc and other heavy elements in red giant stars in particular see [50]).

Within the scenario of stellar nucleosynthesis, only hydrogen and about two thirds of solar helium were made during the prestellar phases of the

---

[1]Here and in the following, Z represents the number of protons in a nucleus, N the number of neutrons and A the atomic mass number, i.e. the total number of nucleons: $A = Z + N$.

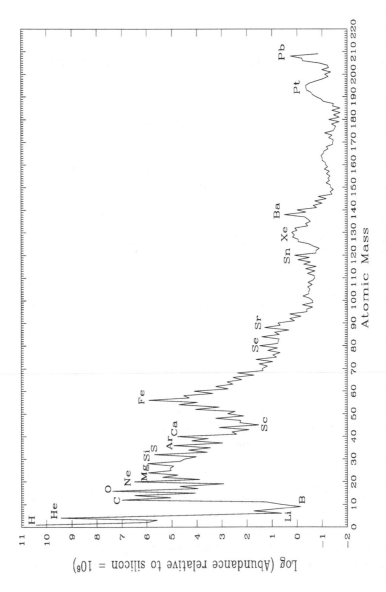

Fig. 2.1 The solar system abundance distribution of nuclear species as derived from observations of the Sun and analysis of primitive chondritic meteorites [20]. For each atomic mass number the isotope of highest abundance is plotted. This distribution, in particular the existence of the several peaks, for example those around Fe and around Ba, must be accounted for by the nuclear processes that produced the elements.

Universe, perhaps together with some of the low-abundance light elements: lithium, beryllium and boron. All the other elements are constantly produced by nuclear processes occurring in stars.

Main constraints to the theories for the origin of the elements are set by the distribution of their abundances in the solar system, which is shown in Fig. 2.1 [20]. This distribution displays several important features, such as the various peaks corresponding to the abundances of several elements, which helped in the identification and early description of the nuclear processes that could be responsible for their production. The theories of nucleosynthesis are intertwined with the theories of nuclear structure as it was soon recognised that the features observed in the solar system distribution of abundances are related to the nuclear properties of each element. For example, the abundances of nuclei around Fe are higher than those of the other elements from S to Pb, which is related to the fact that elements belonging to the "Fe peak" have very stable nuclear structures (see Sec. 2.4).

The basics of the nucleosynthesis occurring in stars were set in 1957 by the work of Burbidge, Burbidge, Fowler and Hoyle [47]. The classification of nucleosynthesis processes into eight types that was proposed by these authors is still mostly valid today and it is summarised in Table 2.1, where each process is associated with its product nuclei and the typical temperature and site where it occurs. Note that in some cases, such as for the r and the p process, it is still much disputed in which astrophysical site these processes occur. The origin of the elements heavier than Fe is still considered one of the greatest unanswered questions of physics [113]. A review article updating the discussion on the eight processes forty years after Burbidge *et al.* has been compiled by Wallerstein *et al.* [287].

Each of the next sections of this chapter is dedicated to one of the processes listed in Table 2.1, with particular attention to the connection with the composition of presolar grains. Each process is described in relation to the stellar environment where it occurs, so that the very basics of current theories of stellar evolution are included in the discussion. Many more details on stellar evolution and nucleosynthesis can be found in the renowned book by Clayton [60].

2.1 Hydrogen burning, and the life of most stars

After the discovery of the nucleus at the beginning of the 1900s, and subsequently of the fact that reactions of fusion and fission among nuclei produce

Table 2.1   The eight types of nucleosynthesis processes.

name	products	T (K)	site
H burning	He, some isotopes of C, N, O, Ne and Na	$> 10^6$	main sequence and red giant stars
He burning	^{12}C and ^{16}O	$\sim 10^8$	red giant stars
α process (C, Ne, O burning)	Ne to S, nuclei with A = integer $\times$ 4	$\sim 10^9$	evolved massive stars, supernovæ
e process (Si burning)	iron-peak species	$\sim 5 \times 10^9$	supernovæ
s process	some elements heavier than Fe: e.g. Sr, Ba, Pb	$\sim 10^8$	red giant stars, evolved massive stars
r process	some elements heavier than Fe: e.g. Xe, Eu, Pt	$\sim 10^9$	supernovæ, neutron stars, ?
p process	p-rich isotopes of the elements heavier than Fe	$\sim 10^9$	supernovæ, ?
x process[a]	D, Li, Be, B		Big Bang, cosmic rays

[a]The x process refers to Big Bang nucleosynthesis as well as interstellar nucleosynthesis due to spallation reactions, when a nucleus is hit by a very high energy particle and smashed into many fragments. This process is not discussed here in detail. For more information see Ref. [287].

energy, it was realised that the fusion of hydrogen nuclei, i.e. single protons (H *burning*), is responsible for the production of nuclear energy in most stars, including the Sun. The discovery that energy is produced by reactions among nuclei actually provided the explanation as to why stars do not collapse under their own gravitational weight [32, 292].

Nuclear reactions are governed by electromagnetic and nuclear forces. The nuclear force is divided into two types, on the basis of their strength: the *strong* and the *weak* interactions. The strong nuclear force only affects quarks, the basic particles composing protons and neutrons. It is responsible for binding quarks together to form protons and neutrons, and also for binding these neutrons and protons together in the nucleus of an atom. Many of the nuclear reactions occurring in stars are governed by the strong nuclear force. The weak nuclear force, on the other hand, affects both quarks and leptons, such as electrons and neutrinos. This force is commonly seen in β decay, a type of radioactive decay in which an electron or positron (β particle) is emitted. Typically, when suffering a β decay, a nucleus changes its charge and is thus transformed into a nucleus of another element. The reaction of fusion of two protons $p + p$ (first line of Table 2.2) is due to the *weak* nuclear interaction: one of the protons is first transformed into a neutron (n) to be fused with the other proton to make deuterium (D = ^{2}H) while a positron (e^+) and a neutrino (ν) are released.

2.1.1 The pp chain

The major chain of nuclear reactions involving proton captures is the *pp* chain, which is illustrated in Table 2.2. The fusion of two protons together by the weak interaction described above starts when the internal region of a cloud of gas collapsing under gravitational force reaches a temperature of about 6 million degrees. In this way stars are born. At such temperatures the gas is completely ionised and all the nuclei are stripped of the electrons that otherwise surround them to make up the atoms. The $p + p$ fusion reaction is activated in the central region of stars like the Sun, which are stars on the *main sequence*, the first and longest phase of their life. The deuterium thus formed captures another proton, producing ^{3}He. At about 8 millions degrees the fusion of two ^{3}He particles can also occur, creating ^{4}He and two protons. Hence, the net outcome of the chain is that four out of every six protons initially entering the chain are transformed into a ^{4}He nucleus[2], on which proton captures do not occur.

Table 2.2   The *pp* chain.

PPI	$p + p \rightarrow D + e^+ + \nu$
	$D + p \rightarrow {}^3He + \gamma$
	${}^3He + {}^3He \rightarrow {}^4He + 2\,p$
	${}^3He + {}^4He \rightarrow {}^7Be + \gamma$
	${}^7Be + e^- \rightarrow {}^7Li + \nu$
PPII	${}^7Li + p \rightarrow {}^8Be$
	${}^8Be \rightarrow 2\,{}^4He$
	${}^7Be + p \rightarrow {}^8B + \gamma$
PPIII	${}^8B \rightarrow {}^8Be + e^+ + \nu$
	${}^8Be \rightarrow 2\,{}^4He$

---

[2]Nuclei of ^{4}He are also known as "α particles" and will be indicated in either way throughout the text.

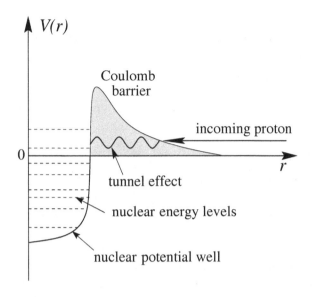

Fig. 2.2 Schematic representation of the shape of the potential energy encountered by a charged particle, e.g. a proton, when approaching a nucleus. During nuclear reactions, charged particles can penetrate the Coulomb barrier thanks to the quantum mechanical tunnel effect, and settle on an quantum energy level of the nucleus thus producing a compound nucleus.

Reactions of proton fusion release energy because the energy stored in a nucleus is smaller than that obtained by simply summing up the energy due to the masses of the nucleons that compose it, each of them contributing an energy $E = mc^2$. This can be understood by picturing the nucleus as a well of negative potential energy: when nucleons fall and are trapped in the well some of their energy is radiated away, similarly to the situation when an electron is captured by an atom and thus releases energy in the form of a photon. Since the mass of a nucleus of ^{4}He is 4.003 atomic mass units, while the mass of a nucleus of H is 1.008 atomic mass units, it turns out that the equivalent energy of a mass of 0.03 atomic mass units (about 30 MeV, see Exercise 3.6.1) is produced every time that four protons are transformed into ^{4}He. Nuclear energy is released in the form of electrons, positrons, high energy photons (γ-rays) and neutrinos. In relation to the shape of the potential energy well around a nucleus, it is interesting to note that an approaching proton feels a barrier produced by the electromagnetic field generated by the nucleus. This barrier can be penetrated and over-taken only thanks to the quantum mechanical *tunnel effect*, by which the wave-function of the approaching particle includes a probability of passing through the barrier (Fig. 2.2).

From the point of view of changes in the isotopic composition of material going through the PPI chain, it is interesting to note that the destruction of the rare isotope ^{3}He occurs at higher temperatures than its production. Hence, the isotopic composition of helium can be modified in stellar environments depending on the temperature at which H burning occurs. This effect has to be considered when studying the isotopic composition of helium measured in presolar SiC grains, as will be discussed in Sec. 4.4.

The chain described above is an extremely common nuclear fusion process in stars because it needs temperatures that are easily reached in stellar interiors and its fuel is the most abundant nucleus in the Universe, i.e. the protons. However, it is by no means the only process occurring in stars involving proton captures. At higher temperatures, around 15 million degrees, the fusion of ^{3}He and ^{4}He can also occur, producing ^{7}Be and leading to the activation of two different branches of the pp chain: PPII and PPIII (Table 2.2). Following a further electron capture or proton capture by the ^{7}Be nucleus, ^{7}Li or ^{8}B are respectively made. Eventually ^{8}Be is produced which decays into two ^{4}He nuclei. The final results of this further H burning is again the production of ^{4}He. All the other nuclei involved in the chain are quickly destroyed by β decays or proton captures.

Among those, only ^{7}Li is a *stable* nucleus – it does not spontaneously decay – and could have been originally present in the star. The initial abundance of this nucleus is typically depleted by proton captures in stellar conditions, unless some mixing mechanism is at work allowing ^{7}Be to travel to the external regions of the star, where the temperature is too low for proton captures to occur, and there capture an electron to produce ^{7}Li. In this case ^{7}Li could actually be enhanced at the stellar surface [51, 241, 242]. This mechanism appears to be at work in some red giant stars, in which Li enhancements are observed (for a recent study see Ref. [56]). Red giant stars have convective envelopes at the base of which proton captures can occur. In convective conditions heat is transported by motion, thus the material is constantly carried from the deep hot regions to the cool surface. In favourable conditions, in red giant stars ^{7}Li is predicted to be produced, rather than destroyed. Li is a rather volatile element and has a very low abundance (see Fig. 2.1), hence it has not been measured yet in presolar grains. However, there may be a chance in the future to carry out measurements of Li in large presolar grains, which would be of much interest in complementing the stellar observations of this element.

2.1.2   *The CNO, NeNa and MgAl cycles*

Hydrogen burning is thus responsible for the production of helium in stars. To explain the origin of elements between He and Si, instead, we cannot rely on proton captures, but we must invoke He burning and the α process. This is because proton captures on ^{4}He do not occur, ^{4}He having an extremely stable structure. In fact, nuclei with atomic mass number equal to five actually do not exist! However, when elements between He and Si are *initially* present in the star, proton captures can very much influence how their abundances, and in particular those of their isotopes, are distributed relative to each other. This is of much importance in the study of presolar grains, which show variations of up to several orders of magnitude in the distributions of isotopes of elements between He and Si. This type of nucleosynthesis can produce large effects on the composition of the material involved, however, as noted above, it needs that the nuclei heavier than He are already present in the star. This is a typical example of *secondary* nucleosynthesis. Nucleosynthesis starting only from H and He originally present in the star is instead defined as *primary*.

Proton-capture reactions involving carbon and heavier nuclei become efficient at temperatures higher than about 15 million degrees. As mentioned above, these reactions can occur when nuclei heavier than helium are present in the star, together with the hydrogen. The fraction of mass composed by elements heavier than He in a star is commonly indicated by Z, and named *metallicity*. For example, about 2% of the material present in the Sun is composed of elements heavier than He. Oxygen is the most abundant, followed by carbon and nitrogen. Then, neon, silicon and iron make up most of the rest of the metallicity, with nuclei heavier than iron representing the least abundant species (see Fig. 2.1). The metallicity and the distribution of the abundances of the elements can vary from star to star, depending on when and where the star was born and on the nuclear processes that occur within it. They can also vary within the same star, as different regions could experience different nucleosynthesis events. Hence, the following general description of proton captures on heavy elements should not be taken as an immutable scheme, because which reactions occur depends on which nuclei are available in the first place.

In the standard description there are three proton-capture chains involving elements from carbon to aluminium, beyond which reaction rates for proton captures become negligible because of the large Coulomb barrier generated by the high number of protons of which the nuclei of the heavier elements are formed. The three chains of proton captures are schematically shown in Fig. 2.3 and they are named after the elements involved: the CNO, the NeNa and the MgAl chains. Usually in the literature these proton-capture chains are indicated as *cycles*, and this name is used hereafter. This is because the chain of reactions typically returns to the nucleus from which the chain had started. However, it should be remembered that depending on the conditions at which the chains are activated, this is not always the case. For example, leakage of material from the NeNa into the MgAl cycle occurs via the ^{23}Na(p,γ)^{24}Mg reaction when the abundance of ^{23}Na is high.

We can distinguish two sub-cycles of the CNO cycle: the CN and the NO cycles. The CN cycle starts to be activated around 15 million degrees with proton captures on ^{12}C and ends with a proton capture on ^{15}N, which releases an α particle and produces ^{12}C again. The alternative channel of proton capture on ^{15}N, which produce a high-energy (γ) photon, is much less likely to occur. The NO cycle starts to be activated around 20 million degrees with proton captures on ^{16}O and ends with a proton capture on ^{17}O, which releases an α particle and produces ^{14}N. Alternately, if the temperature is slightly higher, the cycle ends with proton captures on ^{18}O and ^{19}F, which release an α particle and produce ^{15}N and ^{16}O, respectively. The channel of proton capture on ^{19}F releasing a γ photon, instead, is much less likely to occur, so that also in this case the cycle is closed, at least for temperatures of the order of several millions of degrees and in hydrostatic, as opposed to explosive, conditions.

Proton captures on ^{14}N are less likely to occur than all the other reactions involved in the CNO cycle. Hence, if the chain of reactions is allowed to reach equilibrium conditions, i.e. it is given enough time and enough proton fuel, then the final major result of the cycle is the destruction of carbon and oxygen, which are converted into nitrogen. For example, if material of solar composition, which has C, N and O approximately in the proportions 3 : 1 : 8 is processed by the CNO cycle, the final proportions are changed into 0.01 : 1 : 0.01. Isotopic ratios are also largely modified: an initial solar ^{12}C/^{13}C ratio of 89 is reduced to $\simeq$ 3 and an initial solar ^{14}N/^{15}N ratio of 272 is modified to $\simeq$ 10,000. The final ^{16}O/^{17}O ratio is dependent on the temperature at which the cycle occurs and could vary widely, from 50 to 10,000 (solar = 2660), with temperatures below 30 million degrees favouring the production of ^{17}O. As for other minor isotopes, the abundances of both ^{18}O and ^{19}F are strongly depleted by proton captures. The large abundance of ^{14}N, and the smaller abundances of ^{13}C and ^{17}O, produced

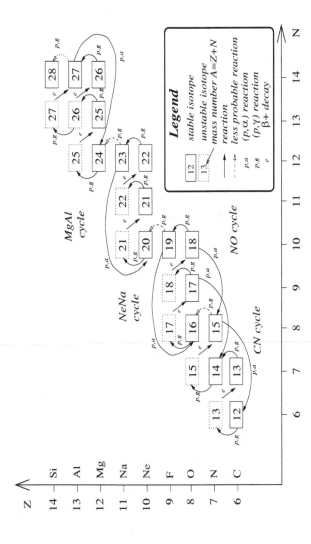

Fig. 2.3   Section of the nuclide chart, in which isotopes are located as function of their number of protons (Z) and number of neutrons (N), schematically displaying the CNO, NeNa and MgAl cycles. Reactions are represented by labelled arrows linking the nucleus entering the reaction to the nucleus produced. Dash-lined arrows represent reactions with typically much smaller probability to occur than the other reaction channels starting from the same nucleus. Label p, g indicates a proton capture with the subsequent release of a γ photon, label p, a indicates a proton capture with a subsequent release of an α particle. Label e indicates that a decay of type β^+ occurs releasing a positron e^+ together with a neutrino. Here, unstable nuclei all decay this way with lifetimes of the order of minutes. The only unstable nucleus that has a lifetime long enough to behave like a stable nucleus and capture a proton is ^{26}Al.

by the CNO cycle are typically of secondary type as they result from the transformation of the ^{12}C and ^{16}O initially present in the star.

The NeNa cycle starts from and returns to ^{20}Ne, whose abundance is nearly left unchanged. At equilibrium, the major result is the production of ^{23}Na at the expenses of ^{22}Ne. Only at temperatures above 50 million degrees is ^{22}Ne reconstituted at the expenses of ^{21}Ne. Similarly, the MgAl cycle starts from and returns to ^{24}Mg, whose abundance is hardly changed for temperatures below 70 million degrees. Since ^{26}Al is an unstable nucleus with the very long half-life of almost one million years, it behaves almost in the same way as a stable nucleus capturing a proton rather than decaying. At equilibrium the major result of the MgAl chain is the production of ^{26}Al at the expenses of ^{25}Mg.

In summary, proton-capture processes typically occur in every star as the necessary temperatures are easily achieved and there is plenty of protons. All presolar grains formed during mass loss episodes around red giant stars, such as the majority of SiC and oxide grains, carry the signature of some sort of processing related to the cycles described above. Their major elemental components, such as C (and also N) in the case of presolar SiC grains, and O and Al in the case of oxide grains show, in fact, isotopic ratios typically affected by proton-capture processes. Also the presence of ^{26}Al in the gas from which the grains formed is recorded in the grains by the abundance of ^{26}Mg, into which ^{26}Al decays. An understanding of the reactions described above, their rates and their occurrence in stars is a necessary ingredient to explain the composition of the majority of presolar grains. In turn, the composition of the grains give us strong constraints in relation to nuclear reaction rates for proton captures and to the modalities in which proton captures occur in stellar interiors, such as their relation to mixing phenomena, the temperatures at which they occur and the initial compositions on which they work.

Finally, note that the operation of the proton-capture cycles as described above is typical for the relatively low temperatures of the order of a few million degrees that are found when considering the evolution of single stars. In some situations, however, proton captures occur at relatively high temperatures, of the order of a few hundreds of million degrees and in explosive conditions. These conditions are related to nucleosynthesis during nova outbursts in binary systems, where two stars orbit around each other. Nova outbursts are thermonuclear runaways occurring when hydrogen is accreted from a companion onto the surface of a white dwarf, i.e. a very compact and dense object with material in degenerate conditions (see Sec. 2.2) left over after the death of stars of low mass. In this case proton-capture reactions are so fast that there is not time for β decays to occur and the resulting nucleosynthesis abundances are very different from those described above, favouring the production of ^{13}C, ^{15}N and ^{17}O. Some presolar SiC and graphite grains show the signature of nova nucleosynthesis and hence the study of this material can set constraints also on this flavour of the proton-capture process.

2.2   Helium burning, and the evolution of stars of low mass

He burning is primarily responsible for the production of ^{12}C and ^{16}O via the two reactions:

$\alpha + \alpha + \alpha \rightarrow\ ^{12}$C, also known as the *triple-α* reaction, and
$\alpha +\ ^{12}$C $\rightarrow\ ^{16}$O $+ \gamma$.

The triple-α reaction requires temperatures of $\simeq 100$ million degrees to occur. First two α particles fuse into ^{8}Be, which however is very unstable decaying back into two α particles within a very short time, of the order of 10^{-15} seconds. Fortunately, the rate of the ^{8}Be$(\alpha, \gamma)^{12}$C reaction is greatly enhanced by the existence of an excited state of the product nucleus, ^{12}C, corresponding to an excitation energy of 7.68 MeV with respect to the ground, stable, state. Thus, a *resonance* in the reaction rate, i.e. a sharp peak in the probability of the reaction to occur, is present at such energy because it is possible for the ^{8}Be $+ \alpha$ system to form a compound nucleus of ^{12}C in such an excited state, which subsequently decays into the stable state of ^{12}C. The existence of the 7.68 MeV state of the ^{12}C nucleus was suggested by Hoyle and subsequently identified by Dunbar *et al.* [86]. Through the triple-α reaction it is possible to overcome the problem of the absence of nuclear structures with $A = 5$, and to produce elements heavier than He.

He burning occurs in the central region of a star after all the protons have been transformed into He. When protons are exhausted in a central region comprising about 10% of the total mass of the star, the star becomes a red giant: the radius increases by up to two orders of magnitude above the initial value, the surface temperature decreases and thus, as for a black-body, the emitted light turns towards the red. The core contracts under its own gravity until the central temperature and density are high enough to allow the triple-α reaction to occur and helium starts being converted into carbon. Carbon can further capture an α-particle and be converted into oxygen. The rate of the $^{12}C(\alpha, \gamma)^{16}O$ reaction governs the relative production of carbon and oxygen and, unfortunately, it is still afflicted by large uncertainties. These uncertainties mainly affect the final composition of the stellar cores that have undergone He burning and of the material ejected by massive stars exploding as supernovæ (see e.g. Ref. [143]).

For temperatures around one billion degrees the reaction chain $^{16}O(\alpha, \gamma)^{20}Ne(\alpha, \gamma)^{24}Mg$ can occur, however, elements heavier than oxygen are more favourably produced by the carbon, neon and oxygen burnings described in the next section.

It has to be noted that, like H burning, He burning can occur in different modes and producing different final results depending on the temperature at which the process occurs and the time-scale of the duration of the process. An important special case is the He burning occurring during the Asymptotic Giant Branch (AGB) phase, which is the evolutionary phase that follows He burning in the core for stars of mass lower than about 8 $M_\odot$. In stars of this mass, after He is exhausted and wholly transformed into C and O, the core contracts to a still higher density. However, the stellar mass is not high enough to compress the central region, and bring it to the high temperatures needed to activate further nuclear burning processes before an electron *degeneracy* develops. For a normal gas, the energy of the particles is defined by the temperature, for a degenerate gas, instead, at increased density, i.e. number of electrons per volume, the electrons are forced to occupy quantum-mechanical states of higher and higher energy. This is because all the possible lower energy levels are already occupied and, because of the Pauli exclusion principle, only two electrons with opposite intrinsic spin can occupy the same energy level. In this way the gas pressure is allowed to increase even for relatively low temperatures and it is possible to balance the gravitational pressure and avoid collapse. Material in a degenerate state is *inert*, in the sense that its pressure does not change following temperature variations.

As shown in Fig. 2.4, during the AGB phase, both He and H burning produce the energy to keep the star in hydrostatic equilibrium. The two burning processes occur in shells above the degenerate core and are active alternately: the H-burning shell is activated most of the time, while the He-burning shell is activated episodically, only for a few hundred years at a time and at temperatures around 200 million degrees. Helium burning, in this case, is only partial and the final result is the conversion of 25% of the initial helium present in the He shell into carbon, while almost no

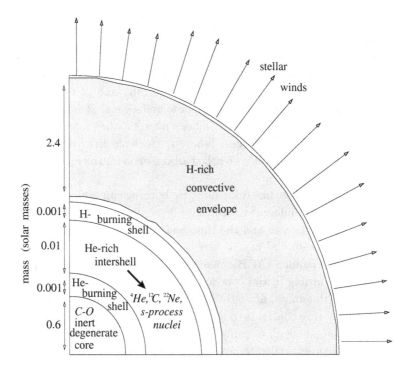

Fig. 2.4   Schematic representation of the structure of an Asymptotic Giant Branch (AGB) star. A rough indication of the mass of each region is given in the vertical axis for a star with initial mass of 3 $M_\odot$. The AGB phase lasts for a few million years and represents an advanced stage of the evolution of stars with masses lower than about 8 $M_\odot$. These stars only experience H and He burning during their lives. Most of the mass of the star is in the convective envelope, which is ejected into the interstellar medium during the AGB phase leading to the formation of a planetary nebula: a hot star surrounded by shells of gas and dust. The degenerate core of the star is eventually left as a cooling white dwarf.

oxygen is produced. Thus, in these stars, He burning is responsible for the production of carbon, rather than oxygen. As will be described in detail in Chapter 4, the occurrence of partial He burning in AGB stars has important consequences for the formation of presolar grains, since AGB stars can become carbon rich and be the site for the formation of carbon dust.

An important secondary chain of reactions is activated during He burning, which involve ^{14}N nuclei as the main seed. A first α-capture reaction on ^{14}N produces ^{18}O, which can be further transformed by a second α-

capture reaction into ^{22}Ne. Since typically He burning acts on the ashes of H burning, and since in the ashes of H burning all initial CNO material has been converted into ^{14}N (see previous section), the typical final result is that after H and He burning all the CNO nuclei initially present in the material are converted into ^{22}Ne.

2.3 The α process: C, Ne and O burnings, and the evolution of stars of high mass

Nuclei made up of α particles (also called α nuclei), i.e. with atomic mass numbers that are multiples of four, are very stable up to ^{32}S, due to the high binding energy of the α particles themselves. This explains why their abundances in the solar system represent some of the peaks of Fig. 2.1. The α nuclei comprise ^{12}C and ^{16}O, which are produced by He burning, and ^{20}Ne, ^{24}Mg, ^{28}Si, and ^{32}S, which are produced by the α process described in this section. Carbon, Ne and O burnings can occur only if the star is more massive than about 8 $M_\odot$ because, as described in the previous section, after He burning in the core, stars of lower mass develop an inert degenerate C-O core. The case is very different for more massive stars, in which the core can be heated up to the high temperatures needed for the α process to occur before having the chance of becoming degenerate. A review on the evolution of massive stars can be found in [297].

Burbidge *et al.* [47] proposed that α-capture reactions on ^{16}O would produce ^{20}Ne, and subsequent α-capture reactions on the resulting ^{20}Ne would produce ^{24}Mg. However, it was later realised that the α-capture reactions on ^{16}O occur at higher temperatures than the main carbon burning reactions:

$$^{12}\text{C} + {}^{12}\text{C} \rightarrow {}^{20}\text{Ne} + \alpha \text{ and}$$
$$^{12}\text{C} + {}^{12}\text{C} \rightarrow {}^{23}\text{Na} + p,$$

which are activated at around one billion degrees. The main C burning reactions thus produce ^{20}Ne and ^{23}Na and release α and p particles. Further α capture on ^{20}Ne and proton captures on ^{23}Na produce ^{24}Mg. Many secondary reactions occur by which all the stable isotopes of Ne, Na, Mg, Al, Si and P can be produced.

At about 1.3 billion degrees the first photodisintegration reaction occurs: ^{20}Ne nuclei are broken back into ^{16}O and α particles by interaction with γ photons: $^{20}\text{Ne} + \gamma \rightarrow {}^{16}\text{O} + \alpha$. The freed α particles interact with the

^{20}Ne producing ^{24}Mg: ^{20}Ne $+ \alpha \rightarrow {}^{24}$Mg $+ \gamma$. This process is known as Ne burning. When Ne is exhausted in the core, the temperature rises again as the core compresses and at about 2 billion degrees oxygen burning, ^{16}O $+ {}^{16}$O, begins. As in the case of carbon burning there are a few main reactions:

$$^{16}\text{O} + {}^{16}\text{O} \rightarrow {}^{28}\text{Si} + \alpha,$$
$$^{16}\text{O} + {}^{16}\text{O} \rightarrow {}^{31}\text{P} + p \text{ and}$$
$$^{16}\text{O} + {}^{16}\text{O} \rightarrow {}^{31}\text{S} + n.$$

These reactions liberate α, p and n particles which interact with the nuclei present, resulting in a large range of products: mainly ^{28}Si and ^{32}S, and also, to various degrees, all the isotopes of Si, Cl, Ar, K, Ca, Ti and Cr.

2.4 The e process: Si burning, and supernova explosions

Si burning starts when the temperature increases above about 3 billion degrees. This process is intrinsically different from H, He, C and O burning as it does not refer to reactions by which two silicon nuclei are fused together, but to a situation in which the high temperature triggers a large number of photodisintegration reactions in particular on ^{28}Si, which is the most abundant product of the previous burnings. The photodisintegration reactions are balanced by extremely fast capture reactions occurring on the large number of freed α, p, and n particles. The material behaves as a gas, composed of nuclei, nucleons, electrons, positrons and photons, trying to reach and maintain a *nuclear statistical equilibrium* (from which we get the name of e process) by means of nuclear reactions. In this process, the formation of the most stable nuclei is favoured, because nuclei that are more tightly bound are the least probable to suffer photodisintegration reactions. The e process hence results in the rearrangement of nucleons into the most stable nuclei. Because the elements belonging to the iron group: Cr, Mn, Fe, Co and Ni have nuclear structures which are the most stable of all elements, the main result of the e process is to convert all the initial material into these nuclei.

States of statistical equilibrium can be represented by considering the entropy of the nuclear gas. The entropy of a system is a function of the temperature and represents the number of allowed ways in which the particles of the system can arrange themselves to share the total energy. By the second law of thermodynamics a system always evolves towards the maximum entropy. This means that it chooses the configuration that can be reproduced in the highest number of ways, and thus is the configuration with the highest probability. When this configuration is reached it is said that the system is in equilibrium. Let us consider this very simple classic picture: a room is separated in two halves by a movable wall in the middle. On one side there is a gas, on the other there is vacuum. If the movable wall is removed, then all the gas particles will start moving around until the density of the gas is constant all over the room. This is because there are far fewer allowed particles configurations to produce a situation in which all the gas particles sit in a corner of the room. On the other hand, there are many more ways to produce the situation in which the gas particles are dispersed in the room. The system evolves towards this case, which has higher allowed configurations, i.e. higher entropy. However, this system is allowed to reach only a given degree of disorder because some constraints are set, such as the fact that the room has solid walls all around.

Also in the case of the *e* process in stars the nuclear gas is not completely free of constraints. One constraint always present is that the number of nucleons must remain constant. This is because conditions in stars are not so extreme to allow nucleons to break down into subnuclear particles, such as free quarks. Other constraints can arise from the fact that there is a time-scale allowed for the process to occur, which may not be long enough for the process to reach complete equilibrium. In such cases some quantities, such as the net abundance of electrons per nucleon and the abundance of nuclei heavier than boron, are set to be constant in the models and the process is described as a *quasi equilibrium* process. In stars various degrees of equilibrium can be achieved by the material during the operation of the *e* process. The maximum entropy achievable, and hence the degree of disorder of the system, can increase or decrease in relation to the number of constraints set by the thermodynamic conditions of occurrence.

The detailed final composition resulting from the *e* process depends on the temperature and density at which the process occurs, as well as on the *neutron excess*. This is defined as the difference between the number of neutrons and the number of protons in the whole material involved in the process (see Exercise 2.6.3). The predominant isotopes of all elements up to Ca have nuclei composed of equal amounts of protons and neutrons: for example ^{4}He has $Z = N = 2$, ^{12}C has $Z = N = 6$, and ^{40}Ca has $Z = N = 20$. In contrast, nuclei composed of a larger number of nucleons typically have stable structures only if their large number of protons are diluted by an even larger number of neutrons, which, unlike protons, do not feel the electromagnetic repulsion. A way to measure the neutron excess is to count the number of electrons per nucleon $Y_e = Z/A$. In fact, the production of heavier neutron-rich nuclei occurs via transformation of protons into neutrons (by β^+ decay or electron capture), and the the number of electrons per nucleon in the material decreases. For the nuclei with $Z = N$ indicated above Y_e is equal to 0.5, while it is smaller for heavier nuclei. For example ^{56}Fe, which is the stable isotope with the highest binding energy has $Z = 26$ and $N = 30$, i.e. $Y_e = 0.46$, ^{209}Bi, which is the heaviest stable isotope has $Z = 83$ and $N = 126$, i.e. $Y_e = 0.40$. Typically, during the time-scale at which the e process occurs, Y_e varies slowly and may be taken to be constant and set by the initial composition of the material.

A neutron excess can be initially present in the material because of the previous nucleosynthesis. In massive stars the main products of the burning episodes that precede the e process are α nuclei which have $Z = N$. However, nuclei present in small amounts deriving from less important nucleosynthetic paths can contribute to a neutron excess. For example, we have observed when discussing He burning that a secondary nucleosynthesis product is ^{22}Ne. This nucleus has $Z = 10$ and $N = 12$, thus $Y_e = 0.45$ and its abundance contributes to a neutron excess in the material. Such neutron excess, corresponding to a decrease of Y_e, is generated by the β^+ decay of ^{18}F into ^{18}O.

High neutron excesses, of the order of 1%, allow the final resulting products of the e process to shift towards nuclei with neutron to proton ratios higher than 1, directly producing ^{54}Fe and ^{56}Fe. This happens when silicon burning occurs in the cores of massive stars. In this situation the system can become very weakly constrained because the time-scale is relatively long and the density is very high. Hence the system can evolve to almost complete disorder and to high neutron excesses. The core of the star evolves to Si burning and reaches a final stage in which Fe and Ni are the predominant elements. Meanwhile, the outer layers of the star experience the various burning phases described above as the temperatures increase in the different locations. A schematic representation of the resulting "onion-like" structure of the star is shown in Fig. 2.5.

As mentioned above, the main products of the e process are the nuclei Fe and Ni which are the elements with the highest binding energy per nucleon. Both fusion and fission reactions involving these nuclei are endothermic rather than exothermic, i.e. they steal energy from the gas rather than give it to it. Hence, when all the core material is turned into Fe and Ni

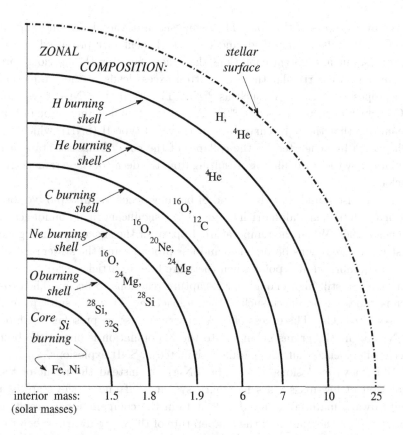

Fig. 2.5 Schematic representation of the "onion-like" structure of a massive star just prior the core collapses resulting in a supernova of type II explosion. A rough indication of the location in mass of each shell in the case of a star with initial mass of 25 $M_\odot$ is given in the horizontal axis. While the core of the star experiences all the major burning phases, the last of which is represented by Si burning, different burning phases are activated in concentric shells around the core depending on the temperature of each region. The major products of each type of burning is indicated.

there are no ways to produce nuclear energy to stop gravitational collapse. Contraction and heating of the core result, leading to partial disintegration of Fe and Ni into α particles and neutrons, and subsequent dynamical collapse of the core. This causes a shock, i.e. a sudden acceleration, to propagate outwards, leading to a supernova explosion. The material in and close to the core is trapped in the final neutron star or black hole. Hence, the iron-peak elements produced in this region are not ejected into the interstellar medium. The supernova shock triggers the explosive burning of the outer layers of the star. The composition of the Si-rich and part of the O-rich shells are modified, and Fe-peak elements are produced by the e process, but in different conditions than those found in the previous core Si burning. In particular the low neutron excess leads to the production of isotopes with $Z = N$, such as ^{56}Ni. The unstable ^{56}Ni decays into ^{56}Co, which in turn decays into ^{56}Fe. An observational confirmation of this scenario is that the light curves of supernovæ of type II (SNII), which are believed to be generated by the collapse of the core of massive stars, are explained by the γ-ray photons resulting from the decay of these radioactive nuclei.

It was also found that the O and Si burning occurring in explosive conditions results in an "alpha-rich freeze out", a significant final abundance of α nuclei [296]. When the temperature drops after the shock wave has gone past nuclear reactions become less and less probable and the system moves out of equilibrium at a point when there are free α particles available at a temperature still high enough for α-capture reactions to occur. Then, α-rich isotopes are produced such as the radioactive ^{44}Ti, which subsequently decays into ^{44}Ca. The excess of ^{44}Ca observed in some presolar graphite, Si$_3$N$_4$ and in SiC grains belonging to the X population is one of the best pieces of evidence that these grains originated in SNII explosions.

Supernova explosions of type Ia (SNIa) are instead the result of the evolution of low-mass stars in binary systems. If one of the stars is a white dwarf, material accreted onto it from the companion can result in the growth of the mass and the temperature of the white dwarf. When the mass comes close to the Chandrasekhar limit mass of 1.4 $M_\odot$ the electron degeneracy pressure cannot sustain the gravitational pressure anymore. In a certain range of initial masses of the white dwarf and of the accretion rates of mass from the companion, this results in carbon being explosively ignited in the centre of the white dwarf and an explosive deflagration or detonation wave, depending on whether the speed of the wave is lower or higher than the speed of sound, propagates outwards causing explosive burning and the disruption of the entire star. The e process occurring in this type of supernovæ represents the major contributor to the abundance of the Fe-peak elements in the Galaxy [274]. In SNIa the nuclear gas is typically allowed to reach a highly disordered state in which the neutron excess can increase to high values and, in some special cases, even produce rare neutron-rich isotopes such as ^{50}Ti [295].

2.5 The production of elements heavier than Fe

Neutron-capture processes are the favoured way for the production of elements heavier than Fe. This is because capture of charged particles, such as protons and α particles, are inhibited by the large Coulomb barrier generated by the high number of protons of which the nuclei of these elements are formed. For the operation of neutron-capture processes it is necessary that a neutron-source reaction is activated, for example the $^{13}C(\alpha, n)^{16}O$ and $^{22}Ne(\alpha,n)^{25}Mg$ reactions. When free neutrons are produced they can decay into protons with a half-life of about 10 minutes or they can be captured by other nuclei. When neutrons are captured by Fe nuclei and their progeny it is possible to produce almost all the heavy elements up to lead (Pb, $Z = 82$) and uranium (U, $Z = 92$). Only a few heavy isotopes of extremely low abundances in the solar system cannot be produced by neutron captures and their abundances are ascribed to proton captures or disintegration reactions (the p process). For a general review see Ref. [190].

As shown in Fig. 2.1 the distribution of the abundances of heavy elements in the solar system decreases to very low values with increasing atomic mass number. The distribution reaches its minimum, of a few 10^{-8} of the Fe abundance, in tantalum (Ta, $A = 181$), the rarest element in the solar system. The solar heavy element distribution is characterised by various peaks which can be explained by the concurrent operation of two neutron-capture processes: *slow* (s) and *rapid* (r) neutron captures. The s process operates in conditions characterised by low neutron densities, of the order of $10^7 - 10^8$ neutrons per cubic centimetre (cm^{-3}). The r-process, on the other hand, operates when the neutron densities are higher than $\simeq 10^{20}$ cm^{-3}. Of course, neutron-capture processes also occur when neutron densities are between 10^{10} and 10^{20} cm^{-3}, with features between the s and the r processes. In spite of appearing as a simplistic idealisation, the classification of neutron-capture processes based on two very different ranges of neutron densities has turned out to be very useful. It provided an explanation for the general features of the solar abundance distribution, the opportunity of producing a clear description of these processes, the chance of creating numerical models to calculate them, and thus the possibility of generating theoretical predictions to be tested against observations. Moreover, it appears that the two processes actually do occur almost in their pure form in different stellar sites, as mainly testified by the anomalous meteoritic Xe-S component (see Sec. 1.2). For the s process, the strongest observational proof of its occurrence in pure form is provided by the *s*-process signature of heavy elements measured in most of SiC grains (Chapter 5) and a few graphite grains (Sec. 6.2). Hence, the distinction between *s* and *r* process has been most successful even if one should always consider the possibility of the occurrence of intermediate situations. This is indicated for example by the *r*-process signature of Xe in diamonds (Sec. 6.1) and the measurements of heavy elements in SiC grains belonging to the X population (see Sec. 4.8.3).

The different paths of production of the *s* and *r* processes through a selection of heavy elements (from yttrium to ruthenium) are shown in Fig. 2.6. Neutron captures proceed through (n, γ) reactions shifting the composition of the material towards neutron-rich nuclei. Because of the relatively low neutron densities, during the *s* process unstable nuclei preferably decay rather than capture another neutron. Hence, neutron captures during the *s* process follow the thick black line in Fig. 2.6 along a path mostly defined by stable nuclei.

For the *r* process the situation is reversed: at high neutron densities unstable nuclei preferably capture another neutron rather than decay. During the neutron flux isotopes with very high numbers of neutrons are produced. These isotopes are extremely unstable against β^- decay and when the neutron flux ends they decay towards the stable nuclei corresponding to their atomic mass numbers. Nuclei with the same atomic number are located in a nuclide chart, such as that presented in Fig. 2.6, along diagonal lines. Hence the dashed arrows labelled as *r* process in the plot represent the material that, from the very neutron-rich region of the nuclide chart, decays towards stable material at the cessation of the neutron flux.

The solar distribution of the abundances produced by the *s* and the *r* processes (see Fig. 2.1) is constrained by the presence of heavy nuclei composed of a *magic number* of neutrons. For elements heavier than Fe, these are the numbers 50, 82 and 126. For lighter elements, the magic numbers are 2, 8, 20 and 28. A nucleus that is composed of a magic number of protons and/or neutrons is very stable. The existence of magic numbers can be explained when nuclei are described as systems ruled by quantum mechanics, in an equivalent way as done for atomic systems. In an atom there are discrete levels of energy allowed for the electrons spinning around the nucleus. Then, there is a given number of electrons allowed to populate each level, depending on how many different values of the angular momentum, the magnetic momentum and the spin are associated with the same energy. When a given energy level is completely filled, then the atom is very stable. As a consequence it is chemically inert, which means that

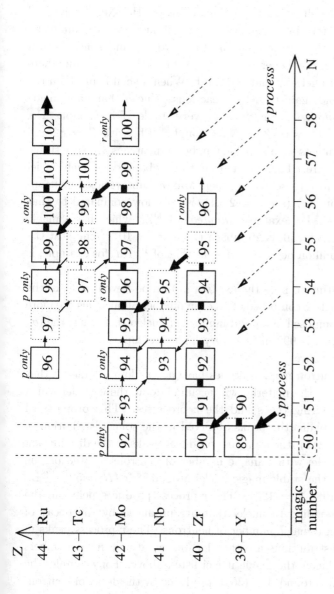

Fig. 2.6 Section of the nuclide chart (similar to Fig. 2.3, where solid- and dotted-lined boxes represent stable and unstable nuclei, respectively) including some isotopes of elements from Y to Ru. The dash-lined arrows represent the r-process path, while the thick black line and the thick arrows represent the chain of (n,γ) and β-decays, respectively, building up the main s-process path (see text for details). Nuclei labelled as s, r or p only are nuclei that can only be produced by the s, r and p processes, respectively. Thin arrows represent the variety of minor paths activated during the s process, leading to the destruction of r- and p-only nuclei. Neutron captures during the s process can occur on unstable nuclei with half-life $T_{1/2}$ longer than the typical time-scale of the s process ($\sim 10^4$ yr), e.g. ^{93}Zr ($T_{1/2} = 1.5$ million yr) and ^{99}Tc (2.1×10^5 yr), which are on the main s-process path.

it does not react easily with other atoms because it does not have any extra electrons or any empty energy level to share: its outer energy *shell* is *closed*. This is the situation for the noble gases: He, Ne, Ar, Kr and Xe. The same interpretation is applied to explain nuclear magic numbers. In the same way as the electrons in an atom, protons and neutrons in a nucleus are allowed discrete energy values, and for each energy value there is a given number of nucleons that can fill it. When a shell is full then the nucleus is very stable against nuclear reactions. Nuclei that have magic numbers of *both protons and neutrons* are very stable. For example ^{4}He, with $Z = N = 2$, ^{16}O, with $Z = N = 8$ and ^{208}Pb, with $Z = 82$ and $N = 126$. Heavy elements with a magic number of neutrons are very stable against neutron capture. This means that the probability of such nuclei to capture a neutron, represented by their Maxwellian-averaged neutron-capture cross section, $\langle \sigma \rangle$ (see Sec. 2.5.1), is very low compared to that of other heavy nuclei. For example, $langle\sigma\rangle$ of ^{90}Zr, which has a magic number of neutrons $N = 50$, is 20 mbarn (1 mbarn $= 10^{-27}$ cm^2), at the temperature of 35 million degrees[3], whereas that of ^{102}Ru, for example, is 192 mbarn.

It is also interesting to note the existence of an odd-even effect in the stability of nuclei. Those composed of an even number of nucleons are more stable than those composed of an odd number of nucleons. In fact, the $\langle \sigma \rangle$ of ^{101}Ru, for example, is 992 mbarn, very high in comparison to those of the nuclei discussed above.

The fact that neutron magic isotopes have a very low $\langle \sigma \rangle$ means that during a neutron flux the material will accumulate at these nuclei which act as bottlenecks during the s and the r processes. This property of neutron captures explains the peaks observed in the solar distribution of heavy elements (Fig. 2.1): the s process produces peaks in the distribution corresponding to nuclei with a magic number of neutrons located on the s-process path, i.e. the stable magic nuclei around ^{88}Sr ($N = 50$), ^{138}Ba ($N = 82$) and ^{208}Pb ($N = 126$). The r process produces peaks in the distribution corresponding to nuclei that are produced by the decay of unstable nuclei with a magic number of neutrons. The r-process peaks in the solar system distributions are thus located at atomic mass numbers approximately ten below the magic nuclei listed above. For example, the r-process peak at Xe, around $A \simeq 130$, is produced by the decay of neutron-rich unstable isotopes of Cd, In and Sn, with magic number of neutrons

---

[3]for historical reasons usually taken as reference temperature in the literature

$N = 82$.

An interesting and helpful characteristic of the heavy element abundances in relation to their production is the fact that some isotopes can only be produced by one single process. These are the isotopes labelled as s, r and p only in Fig. 2.6. Isotopes that can only be produced by the r process are those that do not lie on the s-process path: neutron captures cannot produce them if the neutron density is that typical of the s process because they are preceded, in terms of number of neutrons, by an unstable nucleus that decays rather than capture another neutron. These nuclei lie on the neutron-rich side in the nuclide chart, i.e. at the right of the stable nuclei on the s-process path. Isotopes that can only be produced by the s process are those that lie on the s-process path and are shielded from r-process production by the presence of an r-only isotope with the same atomic mass number. For example, in Fig. 2.6, ^{100}Ru can only be produced by the s process because of the presence of ^{100}Mo: the r-process path represented by the β^- decaying material stops at ^{100}Mo, which is stable, and hence cannot reach ^{100}Ru. A few isotopes such as ^{92}Mo and ^{94}Mo lie on the proton-rich side in the nuclide chart, i.e. at the left of the stable nuclei on the s-process path. They cannot be produced, but are rather destroyed by neutron captures. These isotopes have extremely low abundances compared to the other isotopes of the same elements and are produced by the p process: proton captures or disintegration reactions.

2.5.1 *The s process*

As mentioned at the beginning of this chapter, enhancements of heavy elements typically produced by the s process, such as Sr and Ba, and the presence of radioactive Tc were observed in the 1950s in a class of red giant stars, classified as *chemically peculiar red giants*. These stars are identified as the main site for the s process, producing the *main* and *strong* component of the s-process solar distribution, i.e. s-process isotopes from Sr to Bi ($90 < A < 209$). They are theoretically interpreted to be low-mass stars on the AGB phase briefly discussed in Sec. 2.2. The s process in AGB stars will be discussed in much more detail in Chapter 5 in relation to the origin of SiC grains and their s-process signature. Note that the s process also occurs in evolved massive stars during the hydrostatic He and C burning phases. In this case the *weak* component of the s-process solar distribution is produced, i.e. the s-process isotopes between Fe and Sr ($A < 90$) [159, 228].

During the s process, the abundance N_A of an isotope A along the

s-process path represented in Fig. 2.6 varies with time as:

$$\frac{dN_A}{dt} = production\ term\ -\ destruction\ term$$

$$= N_{A-1}N_n\langle\sigma_{A-1}(v)v\rangle - N_AN_n\langle\sigma_A(v)v\rangle$$

where N_n is the neutron density and

$$\langle\sigma_A(v)v\rangle = \int_0^\infty v\ \sigma_A(v)\ \phi(v)\ dv,$$

indicates the average value over a Maxwell-Boltzmann distribution of velocities $\phi(v)\ dv$ of the product of the relative velocity v times the cross section $\sigma_A(v)$. Typically, for neutron captures it is seen that $\sigma_A(v) \propto 1/v$, thus $\langle\sigma_A(v)v\rangle$ is very nearly constant and it is useful to define the Maxwellian-average $\langle\sigma_A\rangle$ of the cross section $\sigma_A(v)$ so that:

$$\langle\sigma_A\rangle v_{th} = \langle\sigma_A(v)v\rangle,$$

where v_{th} is the thermal velocity[4]. When replacing time with the time-integrated neutron flux, or *neutron exposure* τ:

$$\tau = \int_0^t N_n v_{th} dt$$

one has:

$$\frac{dN_A}{d\tau} = N_{A-1}\langle\sigma_{A-1}\rangle - N_A\langle\sigma_A\rangle$$

which in steady state conditions, i.e. when $\frac{dN_A}{d\tau}$ approaches zero, yields

$$\langle\sigma_A\rangle N_A \simeq constant,$$

which means that, in first approximation, one can derive the relative abundances produced by the s process simply by considering the relative ratios of the neutron-capture cross sections. This "rule of thumb" for the s process, however, is only valid locally because of the presence of the bottlenecks corresponding to neutron magic nuclei. Because of these bottlenecks the value of $\langle\sigma_A\rangle N_A$ is **not** $\simeq constant$ over the whole abundance distribution, but is characterised by steps occurring at the neutron magic nuclei, as shown in Fig. 2.7. The depth of these steps is a function of the neutron exposure, so that different neutron exposures lead to different $\langle\sigma_A\rangle N_A$ distributions. For the highest value of τ shown in the Fig. 2.7, all nuclei up to ^{205}Tl are

---

[4]Note that often in the literature $\langle\sigma_A\rangle$ is simply written as σ_A.

in local equilibrium, as the flux proceeds to the nuclei at the end of the s-process path: ^{208}Pb and ^{209}Bi.

So far we have considered that during the s process unstable nuclei always decay rather than capture another neutron. However, this is an approximation not always valid. For some values of neutron density and temperature, some unstable nuclei with relatively long half-lives can capture neutrons rather than decay. In these cases the s-process path splits into two branches. In Fig. 2.8 the branching paths starting at the unstable nucleus ^{95}Zr, whose half-life is 64 days, are shown. If *branchings* are open, nuclei that otherwise would be classified as r only can be produced by the s process. The *branching factor* indicates the probability that the s-process path will deviate from the standard path. In other words, the branching factor is the probability that the unstable nucleus at the branching point will capture a neutron before decaying. A branching factor is computed as $f_n = \lambda_n/(\lambda_n + \lambda_\beta)$, where $\lambda_n = N_n\langle\sigma(v)v\rangle$, $\sigma(v)$ is the neutron-capture cross section of the unstable nucleus and λ_β is its decay rate, i.e. $\lambda_\beta = ln\,2/T_{1/2}$, where $T_{1/2}$ is the half-life in seconds. Hence, the branching factor is a function of the decay rate of the unstable nucleus, its neutron-capture cross section and the neutron density. In an indirect way, it can also be a function of the temperature when the neutron-capture cross section and/or the decay rate vary with temperature.

The case of the branching point at ^{95}Zr is an important example because the branching factor for the production of ^{96}Zr varies largely with the neutron density. No ^{96}Zr is produced if the neutron density is lower than about 5×10^8 cm^{-3}, but when the neutron density is as high as 10^{10} cm^{-3}, up to a 50% of the s-process flux goes through ^{96}Zr, producing it in an amount dependent on its own cross section. Hence, the abundance of ^{96}Zr is a discriminator of the neutron density at which the s process occurs. The recent measurements of Zr isotopic ratios in single SiC grains are a very valuable constraint for the s process in red giant stars. This will be discussed in detail in Chapter 5.

2.5.2 *The r process*

As introduced earlier, the r process is characterised by very high neutron densities so that nuclei with a very high number of neutrons are produced. However, for each given element, bound neutron-rich isotopes can be be found with a number of neutrons up to a given value, after which it is not possible to capture another neutron because no energy states exist for which

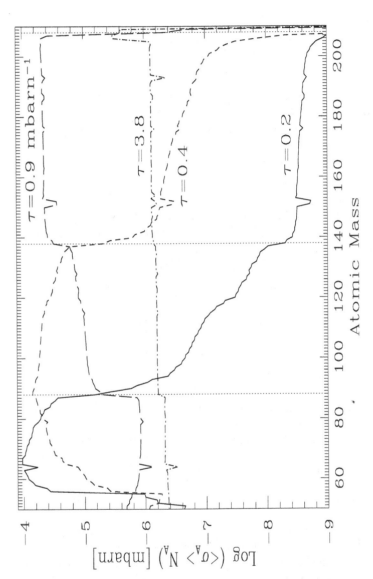

Fig. 2.7 The $\langle \sigma_A \rangle N_A$ distribution produced by the *s* process after different neutron exposures as a function of the atomic mass *A*. Neutron-capture cross section values are chosen for a temperature of 8 keV, corresponding to $\simeq 9 \times 10^7$ K. Different distributions are obtained for the different values of the neutron exposure τ, as indicated. Dotted lines are drawn in correspondence to the neutron magic nuclei 88Sr, 138Ba, and 208Pb. Local equilibrium $\langle \sigma_A \rangle N_A \simeq constant$ is roughly verified far from nuclei with magic neutron numbers. Higher neutron exposures produce heavier nuclei as the bottlenecks at neutron magic numbers are more easily bypassed.

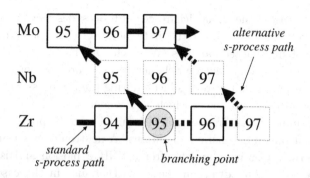

Fig. 2.8 Section of the nuclide chart showing the branching path that is open at ^{95}Zr in conditions of high neutron density. When the branching is open, part of the flux goes through ^{96}Zr, skipping ^{95}Mo and ^{96}Mo.

the neutron could be bound to the nucleus. This is called the *neutron drip* point. During the *r* process, when nuclei with numbers of neutrons at the neutron drip point are abundant, an equilibrium between neutron-capture (n, γ) and inverse photodisintegration (γ, n) reactions is established. At that point β^- decay reactions also start playing a role since isotopes with high numbers of neutrons β^- decay very quickly, with half-lives typically of the order of seconds. Models of the *r* process have to include a very large number of unstable isotopes. The properties of these nuclei, such as neutron-capture cross sections and β-decay rates are difficult to measure and are usually calculated by theoretical nuclear models.

For the *r* process to produce the solar *r*-process abundance distribution of elements up to uranium, with $A = 238$, it is necessary to have a very large number of free neutrons. The typical seed nuclei, i.e. the nuclei on which the neutron-capture process starts, are nuclei belonging to the iron peak and have A around 60 to 100. This means that the *r* process is efficient enough to produce the solar abundances of *r*-process elements if there are about 100 to 200 free neutrons present per seed nucleus. The exact composition of the seed nuclei depends on the previous nucleosynthesis phases, so that the exact number of neutrons needed to produce the solar distribution depends on the site where the *r* process occurs.

The modelling of the *r* process is challenging not only in the nuclear physics involved, but also with regards to the site where it could occur. For a detailed review of what is described below see Ref. [190]. Possible sites for the *r* process are SNII explosions, in particular at the time when, as the temperature drops, material freezes out from nuclear statistical equilibrium (similar to the case of the alpha-rich freeze out process described in Sec. 2.4). When the nuclear gas cools down the first reactions to become unimportant are those involving charged particles. When charged-particle reactions stop and if the material is very neutron rich, neutron-capture reactions can occur on seed heavy nuclei that were produced during the e process.

There are other situations in which a very neutron-rich material exists, for example, in neutron stars, neutron-rich nuclei could be present together with free neutrons in degenerate conditions. A neutron star is usually the core of a former massive star left after a SNII explosion and is similar in structure to a white dwarf in the sense that also in the case of the neutron star the pressure produced by degenerate material sustains the star against gravitational collapse. In the case of the neutron star, though, the density approaches the density of the nucleus and the degeneracy is produced by neutron pressure rather than by electron pressure. If material from a neutron star escapes into space then it could be the site for another flavour of the r process, this time unrelated to statistical equilibrium.

Various scenarios have been proposed for the r process in relation to SNII explosions and neutron stars, with possible locations varying from the region at the boundary of an exploding massive star core, where material escapes into space or falls into the neutron star; rotating stellar cores with or without magnetic fields; the disruption of a neutron star turning into a black hole because of accreting material from a companion; and neutron star-neutron star collisions. In all these cases the r-process products are of primary nature, i.e. the production of the seed nuclei from which the building of heavy elements starts occurs in the same site as the production of heavy elements.

There are also situations in which the r-process products are of secondary nature, i.e. the seed nuclei were initially present in the stellar site where the production of heavy elements occurs. In these cases the r process is unrelated to statistical equilibrium and the neutrons are produced by α-capture reactions such as $^{13}C(\alpha, n)^{16}O$, $^{22}Ne(\alpha, n)^{25}Mg$ and $^{25}Mg(\alpha, n)^{28}Si$, which occur at temperatures lower than about one billion degrees, at which statistical equilibrium has not set in. Large numbers of neutrons can be released by the reactions listed above, for example, in the helium shell of an exploding SNII. This model does not seem capable of making enough neutrons to reproduce the solar system r-process distribution. However, it can be used to explain the isotopic composition of heavy elements observed in presolar grains that originated in SNII explosions, such as the Xe-H component carried by presolar diamonds (Sec. 6.1), as well as the composition of Mo and Zr measured in SiC grains of type X (Sec. 4.8.3) and in some graphite grains (Sec. 6.2). These compositions, in fact, do not display the signature of a proper r process, but rather that of a weak neutron burst. Also, the detailed modelling of nucleosynthesis in the He shell during a SNII explosion, led to the discovery of another type of process: the ν process. During the SNII event spallation reactions breaking down nuclei in the helium shell because of the bombardment of neutrinos from the nascent neutron star, result in the ejection of nucleons. In this case an interesting type of nucleosynthesis results leading, for example, to the production of the rare element fluorine.

All the scenarios described above present unresolved problems in trying to match the solar r-process abundance distribution and to explain other observational constraints related to the r process. These constraints include:

(1) experimental and theoretical nuclear properties, which, together with the solar r-process abundance distribution, indicate that the r process did reach the $(n, \gamma) - (\gamma, n)$ equilibrium conditions, and that the three r-process peaks in solar distribution were produced by different components [158];

(2) observations of r-process elements in stars of very low metallicity, which point to a primary origin for the r process and support the idea that different r-process production sites are responsible for lighter and heavier r-process elements (see e.g. Ref. [256]);

(3) the total abundance of r-process elements compared with the rate of occurrence of SNII. This comparison indicates that SNII would produce too much r-process material because a large fraction of the core must be expelled in order to recover the elements produced by the r process, given that the needed high neutron excesses are only found deep in the core.

A possible site for the r process has been identified in the neutrino-driven winds of a neutron star that is forming just after a SNII explosion (see Ref. [190] for a discussion). However, the potential of this site for the r process remains an open question and many studies have been devoted to provide an improved physical description of this situation, for example considering the role of magnetic fields [276], the effects of boundary conditions [273] and the inclusion of light neutron-rich nuclei [272]. Other more or less exotic alternative scenarios have been proposed, such as r-process nucleosynthesis in relation to the possibility of *prompt* explosion in SNII, i.e. when the shock wave itself rather than absorption of neutrinos from the core is the mechanism responsible for ejecting the mass of the star [268], the production of r-process elements in accretion disks around newly formed neutron stars and their ejection by jets from the disks [52], the possibility of having r-process nucleosynthesis without a neutron excess [191], and the accretion-induced collapse of a white dwarf into a neutron star in a binary system [227]. The mechanism for the production of the r process elements is still elusive and remains one of the greatest puzzles of modern astrophysics [113].

2.5.3   *The p process*

As shown in Fig. 2.6, p-only nuclei cannot be produced by neutron captures. These nuclei have very low abundances in the solar system with respect to the other stable nuclei of the same elements. For example the abundances of the p-only nuclei ^{136}Ce and ^{138}Ce represent only 0.19% and 0.25%, respectively, of Ce in the Solar System. However, there are a few p-only nuclei that are exceptions to this rule: ^{92}Mo and ^{94}Mo, which represents 15% and 9%, respectively, of the total abundance of molybdenum, and ^{96}Ru and ^{144}Sm, which represents 5% of the total abundance of ruthenium and samarium, respectively.

The production of p-only nuclei can be ascribed to proton captures or to disintegration processes. In both cases these nuclei are produced starting from material already enriched in heavy elements. For proton captures, the scenario could be similar to that applicable to the r process in relation to nuclear statistical equilibrium: in the case of the r process the material freezing out from equilibrium must be neutron rich, in the case of the p process the material must be proton rich. In this situation the production of proton-rich nuclei by proton captures is balanced by disintegration reactions (γ, n) and β^+ decays of unstable nuclei. In the alternative scenario, proton-rich nuclei are produced by the disintegration of heavier elements. Within the disintegration scenario it is possible in principle to explain why ^{92}Mo and ^{144}Sm are more abundant with respect to other p-only nuclei: these two p-only isotopes have a magic number of neutrons ($N = 52$ and 80, respectively) hence their disintegration (γ, n) rates are small and they act as bottlenecks when disintegration processes are at work. Some p-only isotopes can also be produced by the ν process described in section 2.5.2. As in the case of the r process, laboratory β-decay rate measurements are crucial to the modelling of the p process [247].

It is reasonable to assume that the production of p-only nuclei occurs in different sites, such as SNII explosions [230], supernova-driven accretion disks around neutron stars [96] and also SNIa explosions [103]. A major constraint to the models comes from the solar abundance distribution, in particular the high abundance of ^{92}Mo. Constraints also come from presolar material: the presence of extinct radioactive ^{146}Sm, a long-living isotope produced only by the p process, has been discovered in meteorites [225]. In the Xe-HL component of presolar diamonds the signature of the p process appears together with that of the r process (Fig. 1.3). However, no detailed studies have been recently dedicated to the explanation of such a signature. Also one presolar SiC grain of type A+B has been discovered showing the signature of the p process in its Mo composition [244] (Sec. 4.8.2).

2.6 Exercises

(1) The relative abundance of each nuclear species i in a given material is usually represented by the fraction of the mass of the species i with respect to the total mass: $X_i = M_i/M$, or by the abundance in number per atomic mass unit (m_u): $Y_i/m_u = n_i/\rho$, where n_i is the number of particle per unit of volume and ρ is the density. Note that $\sum_i X_i = 1$ by definition. Show that for each species i:

$$Y_i = X_i/A_i$$

where A_i is the atomic mass of species i.

(2) The rate at which a nuclear reaction involving the fusion of nuclei of two different species i and j in a material is calculated as the number of reactions that occur when a flux of nuclei i is moving with a Maxwellian distribution of velocities with respect to nuclei j:

$$n_i n_j \langle \sigma v \rangle,$$

where n_i and n_j are the the number of particle per unit of volume of each species and $\langle \sigma v \rangle$ is the Maxwellian-averaged cross section. The cross section σ represents an area around each target nucleus in which interaction results in fusion. Show that the reaction rate units are number of reactions per unit of volume per unit of time. (Note that to describe reaction rates independent of their site of occurrence often the formula $N_A \langle \sigma v \rangle$ is used, where N_A is Avogadro's number and the

units are number of reactions per unit of volume per unit of time per mole, see e.g. Fig. 5.1).
Express the rate using the quantities Y and X for the two species, as defined in the previous exercise.

(3) The neutron excess of a given material can be calculated as

$$\eta = \sum_i (N_i - Z_i) Y_i,$$

where N_i and Z_i are the numbers of neutrons and protons of each nuclear species, and Y_i is its abundance in number. Show that the neutron excess can also be represented by $1 - 2Y_e$, where Y_e is the **total** abundance of electrons per nucleon.
Starting with a material of typically solar composition, i.e. metallicity $Z = 0.02$, show that after H and He burning the neutron excess of the material increases by 0.0018.

(4) Xenon (Xe, $Z = 54$) has nine stable isotopes with atomic mass numbers: 124, 126, 128, 129, 130, 131, 132, 134 and 136. In the periodic table of elements iodine (I, $Z = 53$) and tellurium (Te, $Z = 52$) precede xenon. Iodine has only one stable isotope: ^{127}I, while Te has eight stable isotopes: 120, 122, 123, 124, 125, 126, 128 and 130.
Draw a schematic nuclide chart similar to that shown in Fig. 2.6 and locate the stable isotopes of Te, I and Xe. Draw the s-process path in this region of the nuclide chart and identify which of the Xe isotopes are s, r and p only. Discuss your findings in relation to the isotopic signatures in presolar grains Xe-S and Xe-HL shown in Fig. 1.3.

(5) Calculate the branching factor for the branching point at ^{95}Zr for $N_n = 5 \times 10^7, 5 \times 10^8, 5 \times 10^9$, and 5×10^{10} cm^{-3}. The half-life of ^{95}Zr does not vary with temperature and is equal to $T_{1/2} = 64$ days. The Maxwellian average neutron-capture cross section of ^{95}Zr at a temperature corresponding to 23 keV is $\langle \sigma \rangle = 60$ mbarn [280], and $v_{th}(23 \text{ keV}) = 2.1 \times 10^8$ cm/s.

Chapter 3

Laboratory Analysis of Presolar Grains

This chapter describes the main laboratory techniques currently used to recover presolar grains from meteoritic rocks, and to analyse them for their elemental and isotopic compositions.

Different procedures have been elaborated to extract or locate the grains, depending on their types. Diamond, graphite and SiC grains are extracted using an isolation (i.e. separation) procedure that involves chemical and physical methods. Other types of grains, such as corundum, have been located and analysed by ion-imaging techniques.

Grains can be viewed using electron microscopes, and their morphologic and mineralogical properties can be determined. Their elemental and isotopic compositions can be obtained by extracting and counting ions of the elements and isotopes that compose them.

3.1 The isolation of diamond, graphite and SiC grains

To recover diamond, graphite and SiC grains physical separation procedures alone do not work because the grains are strongly attached to other components of the matrix of the meteorite such as clay minerals and kerogen, i.e. macromolecular organic carbon. Hence, the various components of the rock need to be destroyed with appropriate chemical solvents. The exact isolation procedure was developed by trial and error, because the nature of the grains was unknown and hence researchers had to be very cautious not to use a chemical agent that would destroy the presolar material. The anomalous noble-gas components (described in Sec. 1.2) were used as the thread to their presolar carriers: at each step it was necessary to check that such components, and hence the presolar material, were still present in the residual material left after the treatment [270].

Hereafter it is summarised the isolation procedure to extract presolar diamond, graphite and SiC grains from a portion (named K) of the Murchison meteorite presented in Ref. [12]. By this procedure it is possible to recover more than 70% of the original presolar grains, as deduced by comparing the amounts of the anomalous noble-gas components at various phases of the treatment. The final result consists of several meteoritic residues in which the different types of presolar grains are concentrated. Depending on the separation procedure, the mineralogical purity of these residues can be very high, e.g. close to 100% for diamonds and higher than 90% for SiC grains. All the SiC grains are of presolar origin, however the fraction of diamonds of presolar origin is still uncertain, as discussed in Sec. 1.4 and Sec. 6.1.

First, Murchison K was disaggregated by freeze-thaw cycles. The material was then separated into fractions with different sizes, from 1 to 1000 μm. The fractions with sizes between 30 and 1000 μm were renamed Murchison K and further processed for the extraction of presolar grains. The laborious steps of the procedure are schematically summarised in Fig. 3.1. The first step involved dissolving the silicates, which represent $\simeq$ 96% the rock. To this aim the material was treated in different cycles using strong acids such as hydrochloric (HCl) and hydrofluoric (HF) acids. The residue, named KA, was composed of mud and contained sticky kerogen as well as sulfur. These were removed in a second step using the strong base KOH (potassium hydroxide), leading to the separate named KB. Oxidation with H_2O_2 (hydrogen peroxide) helped to destroy sulfur compounds and some of the kerogen. At this point nearly 99% of the meteorite was dissolved. The removal of KOH and H_2O_2 were performed by various means, such as ultrasonication, i.e. subjecting the sample to ultrasound so as to fragment the macromolecules in it, decantation, i.e. carefully pouring the solution from a container, leaving the sediments in the bottom of the container, and centrifugation, i.e. applying a centrifugal force so as to more rapidly and completely cause the sediments to settle. At this point a black substance consisting of ultramicroscopic particles (colloid) formed in the material and was extracted. This colloid contained the presolar nanodiamonds.

The sediment left after the colloid extraction, named KC, contained oxides, SiC and other carbonaceous phases from kerogen to graphite. Because graphite has a lower density ($<$ 2.2 g/cm^3) than SiC and spinel ($>$ 3 g/cm^3), it could be extracted by sorting the mixture of minerals by density. Density separation is performed by placing the sediment in a liquid with a given density, in which grains with densities less than that of the liquid float and grains with densities greater than the liquid sink. Sodium

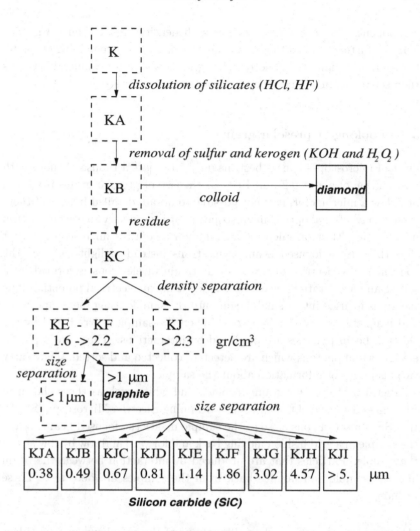

Fig. 3.1 Schematic summary of the steps required for the isolation of diamond, graphite and SiC grains. The average sizes of the different fraction of SiC grains are indicated in the bottom boxes.

polytungstate dissolved in water was used as the high-density liquid. When analysed for the anomalous noble-gas components the fraction with density from 1.6 to 2.2 g/cm^3, KE and KF, turned out to contain 70% of all the Ne-E(L) component. The fractions in such density range were then separated by size with grains lower or higher than 1 μm. The grains with size larger than 1 μm contained presolar graphite, carrier of the Ne-E(L) component. The remaining fraction with density higher than 2.3 g/cm^3, KJ, was further treated to remove spinel as well as any remaining graphite, kerogen, and diamond. The remaining greenish-white residue of SiC was then separated into nine groups of grains of different size.

3.2   Looking at presolar grains

Optical microscopes can at best magnify an object a thousand times with a resolution of about 0.2 μm. Because the size of presolar grains is around or below a micrometer, it is not possible to obtain detailed images of them with a conventional optical microscope and it is necessary to use an electron microscope. Electron microscopes function as optical microscopes except that they use a focused beam of electrons instead of light to "see" the specimen. In electron microscopes a stream of electrons is formed by a source and accelerated toward the sample using an electrical potential. The stream is focused into a small beam onto the sample using metal apertures and magnetic lenses as collimators. When the sample is irradiated by the electron beam particles are produced by interactions. These are typically electrons and photons which are detected, collected and sorted. They carry various pieces of information about the sample.

Pictures of presolar grains in black and white such as those shown in Fig. 1.5 and 1.6 are obtained using a Scanning Electron Microscope (SEM). In a SEM a set of coils close to the sample makes the focused beam "scan" the sample in a grid fashion moving back and forward and row by row across the sample. Particles that are produced by interaction between the beam and the sample and bounce back from the samples are detected. These include:

(1) secondary electrons, released by atoms of the sample when an incident electron passes nearby. These yield information on what the sample looks like so that images can be produced, which are useful to see the surface structure of the grains;

(2) back-scattered electrons, which give information on the number of protons characterising the nuclei of the analysed material; and

(3) X-rays, emitted by the sample when higher-energy electrons fill lower-energy states (typically the K shell for elements such as Mg and Al, and the L and M shells for heavier elements) freed by a small fraction of the secondary electrons. Their energy yields information on the elements of

which the sample is composed so that the chemical nature of the grains can be determined by Energy-Dispersive X-ray spectroscopy (EDS).

By SEM X-ray mapping of polished sections of meteorites it was possible to identify for the first time presolar SiC grains *in situ* [4], once their chemical nature was a priori known through the grain isolation work.

Also the Transmission Electron Microscope (TEM) is used to "see" presolar grains, in particular to study the structure of the crystals on very small scale. In this case the sample is irradiated by the electron beam and the electrons that pass through the sample are focused to strike a phosphor screen, producing light and generating a visual image. The darker and lighter areas of the image represent, respectively, thicker and denser or thinner and less dense areas of the sample because the thickness and the density of the material determine the number of electrons that are transmitted. Also electron diffraction studies can be carried out with TEM. They provide information on the crystal structure of samples and the identification of specific minerals.

For viewing by TEM, specimens have to be cut into slices thin enough to transmit the electrons. For example if the electrons in the beam have an energy of 200 keV, the specimen must have a thicknesses less than 100 nm. A way of slicing presolar grains is by using a diamond knife attached to an *ultramicrotome*. A single presolar grain is placed at the bottom of a gelatin capsule and then covered with a low-viscosity resin that can become very hard. The gelatin capsule is then removed and a square plateau containing the grain is carved in relief from the resin using a sharp glass knife. The resin block is then mounted into the ultramicrotome chuck that can be moved in increments as small as a few tens of nanometers. The plateau containing the grain is then sectioned with a highly sharpened diamond, and the thin slices float off onto a water surface where they can later be retrieved for study in the TEM.

The use of electron microscopes allows us to investigate the mineralogical and microstructural properties of presolar grains, thus providing invaluable information on the condensation history of the grains [29] (Sec. 1.5.2).

3.3   Isotopic measurements with mass spectrometers

The isotopic composition of a sample is measured by extracting and counting the number of atoms of a given isotope, i.e. the number of atoms of a given mass, since different isotopes have different atomic mass. Mass Spectrometers (MSs) are instruments designed to extract ions from a sample, separate them depending on their mass and measure the number of ions of each mass. The mass resolution of a mass spectrometer is defined as $R = m/\Delta m$, where m is the mass of the ion and Δm is the difference with the next observable mass. Because typically both ions of atoms and of molecules are extracted from a sample, a high MS resolution is important to measure isotopic compositions of some elements by ensuring that atoms are separated from molecules with very similar mass. A high mass resolution is achieved during an experiment by different procedures applied before MS analysis, such as chemical separation and the choice of specific ion sources to eliminate molecular interferences. To prevent any interaction of the ions with the residual gas in the MS, it is necessary to operate MSs under as close to high vacuum conditions as possible, which are produced using vacuum pumps.

Ions of different masses are separated in space or time using different methods. Separation in space is performed by applying and varying an electric and/or a magnetic field. This procedure is based on the fact that ions of different masses follow different trajectories in an electromagnetic field. For example, ion paths are deflected by the presence of a magnetic field and turned into circular orbits in which the radius r is given by the balance between the centripetal force produced by the magnetic field and the centrifugal force so that:

$$r = \frac{mv}{qB},$$

where m is the mass of the ion, v its velocity, q its charge and B is the applied magnetic field. Ions travelling with a given orbital radius can then be selected by an appropriately positioned detector. Separation in time can be performed by means of a time-of-flight MS, which uses the differences in transit time of ions of different masses and the same energy in an electric field. If an electric field with voltage V accelerates ions of charge q into a tube, each ion will have a fixed kinetic energy of

$$qV = \frac{1}{2}mv^2.$$

Hence lighter ions have a higher velocity than heavier ions and reach the detector at the end of the tube earlier. This system operates in a pulsed mode so ions are produced and extracted in pulses.

Once the ions are separated their number is analysed using an ion detector, such as a Faraday cup, which is a metal cup placed in the path of the charge ion beam and attached to an electrometer measuring the resulting electrical current, or an electron multiplier tube or a channel plate, which are designed to eject secondary electrons when struck by an ion hence multiplying the ion current and amplifying the electric signal to produce a detectable pulse for every ion arrival, in this case, counted one by one.

To extract ions from a sample different methods can used, which are explained below.

3.3.1 *Noble-gas extraction*

Noble gases are very difficult to ionise and they are extracted by stepped-heating combustion of the sample to high temperatures, up to 2000 degrees [168]. The temperature of the sample is increased in discrete steps and the gases releases at each step are separated and analysed. Because noble gases have very low abundances in presolar grains, it is necessary to perform measurements on a considerable amount of material so that average information on the composition of many grains is obtained. It is still possible in this case to have differential information by means of considering the results obtained at the different temperature steps.

Isotopic measurements of He and Ne have been possible in single grains through a technique in which the gases are released from the grains using a laser, and then ions are produced by electron impact in the gas [157, 201, 202].

3.3.2 *Secondary Ion Mass Spectrometry (SIMS)*

To measure the composition of single presolar grains it is necessary to extract ions from samples made of very tiny particles. A Secondary Ion Mass Spectrometer (SIMS), also called an *ion microprobe*, allows the analysis of samples of sizes down to ~ 1 μm. In a SIMS, ions are extracted by using a primary ion beam, made for example of Cs^+ ions, accelerated to an energy of $\simeq 20$ keV and focused onto the surface of the sample. A schematic representation of this process is shown in Fig. 3.2.

The interaction of the ion beam and the surface of the sample has three effects: the upper layers of the sample are mixed and turned into amorphous material, atoms from the primary ion beam are implanted in the sample, and some secondary particles (atoms and small molecules) are ejected ("sputtered") from the sample. The ionisation efficiency indicates how much of the sputtered material comes off as ions and it varies depending on the element. For some elements it is as high as 10%, for some other elements it is lower by several orders of magnitude. The ions are extracted from the sputtering area by applying an electric field between the sample and an extraction lens. The lens focuses the ions into a secondary beam, which is finally sent to a mass spectrometer to be analysed.

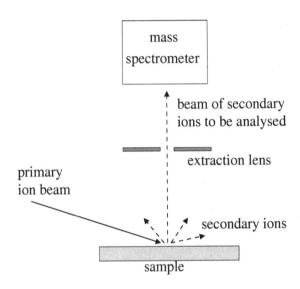

Fig. 3.2 Schematic representation of the ion extraction method in SIMS.

The size of the sputtered area depends only on the diameter of the primary ion beam, which is typically of the order of a micrometer, hence SIMS analysis has a relatively high spatial resolution, allowing the analysis of presolar grains belonging to the largest-size region of the distribution. Also, since material can be continually sputtered from a surface, it is possible with SIMS to determine the composition as a function of distance from the original surface (depth profiles) hence performing measurements in three dimensions. The SIMS technique has been used extensively to measure the composition of SiC and graphite grains in particular at the Laboratory for Space Science of Washington University in St Louis (USA) and at the Max-Planck-Institute for Chemistry in Mainz (Germany).

However, only a small fraction of presolar grains could be analysed, i.e.

Fig. 3.3   Photo of the NanoSIMS at the Laboratory for Space Science of Washington University in St Louis. The main parts of the instrument are indicated (courtesy Frank Stadermann).

grains of sizes larger than about a micrometer (such as SiC grains in the residues KJE, KJF, KJG, KJH and KJI shown in Fig. 3.1). Yet, smaller grains actually constitute the majority of presolar dust and an improved instrument was necessary to analyse grains with sizes down to ~ 0.1 μm (such as SiC grains in KJA, KJB, KJC and KJD). Since 2000 the two laboratories mentioned above have been equipped with the next generation SIMS: the NanoSIMS (manufactured by Cameca, France), whose primary ion beam has a diameter as low as 30 nanometers, as opposed to a few μm for the old instrument (see e.g. Ref. [121]). A photo of the NanoSIMS is shown in Fig. 3.3 and the main parts of the instrument are indicated. The NanoSIMS has an extremely high spatial resolution and it is able to focus on particles with sizes smaller than a micrometer, making it the ideal instrument for recovering and analysing small presolar grains. There are also other improvements over the previous generation of SIMS. The secondary ion transmission efficiency at high mass resolution, necessary for most isotopic analyses, is much higher, at higher mass resolution, than in the old ion microprobes, which means that more secondary ions reach the mass analyser and measurements result in higher precision. Moreover, it is possible to measure different masses simultaneously using five electron multipliers. The NanoSIMS is a new and very complex instrument. There are more than 200 parameters to adjust for each measurement. Not surprisingly it took some time for the instruments to become operational. New discoveries have began to be reported (see Sec. 1.4 and, e.g., Refs. [11, 180, 200, 261, 304]) and it is foreseen that the use of the NanoSIMS will produce considerable progress in the laboratory study of stardust in the coming decades.

3.3.3   The advent of Resonant Ionization Mass Spectrometry (RIMS) in trace element analysis

Great advances have been made in recent years in the task of measuring the isotopic composition of elements heavier than Fe in presolar grains. The problem with such elements is their very low abundance, down to levels of one part per million. They are considered *trace* elements in presolar grains. Previously, Thermal Ionization Mass Spectrometry (TIMS, as well as SIMS [305]) was applied to measure the composition of refractory materials in presolar grains, including heavy trace elements [218, 224, 226, 234]. In this technique, a sample is deposited on a metal ribbon and an electric current heats the metal to a high temperature so that positive ions of different elements at different temperatures are generated by the process of thermal ionization. Because this technique sometimes involves chemical preparation to purify the sample, and because the abundance of trace elements is low, it was necessary to have a considerable amount of material in the sample. Hence it was possible to make measurements only on samples of grains in bulk, i.e. collections of a large number of grains. These measurements, performed on SiC grains in bulk, have yielded precious information on the isotopic composition of many s-process elements, providing strong constraints to the theoretical models (see e.g. Refs. [98, 100]). However, the information is restricted to the average composition of many grains. Differential information is only possible by the analysis of SiC grains of given sizes in bulk. However, the analysis of the elemental and isotopic composition of the grains should be, ideally, performed on single grains, since each grain can in principle have a different composition and add a new constraint to theoretical models.

To measure the composition of trace elements in single grains, instruments of very high sensitivity are needed, to detect as much as possible of the low concentration of such elements. With the SIMS instrument it is possible to measure trace element concentrations in single grains [10], but typically not their isotopic composition because the sensitivity of the instrument is not high enough. In other words, the abundance of ions of trace elements extracted from a single presolar grain is too small to produce a statistically significant signal. With the advent of the NanoSIMS, it is now possible to measure the composition of heavy elements in single grains using the SIMS technique [180, 181]. However, only the composition of some heavy elements can be measured because, for many elements of interest, interferences by isotopes of the same mass (isobars) are present. This means that it is not possible to distinguish between isobars of different elements, such as ^{96}Mo and ^{96}Zr (see Fig. 2.6).

In the mid-1990s a new instrument of high sensitivity (the Chicago-Argonne Resonant Ionization Spectrometer for Mass Analysis, CHARISMA) was developed at the Argonne National Laboratory (USA) with which it is possible to analyse the composition of trace elements in single presolar grains of relatively large size (such as KJG SiC grains, refer to Fig. 3.1) avoiding mass interferences [178, 243]. This instrument makes use of Resonant Ionization Mass Spectrometry (RIMS). The ion production method is schematically represented in Fig. 3.4. One or more lasers (beams of photons all with the same frequency and phase) are tuned to the same energies needed to excite an atom of a given element to higher and higher energy levels, i.e. the lasers are in *resonance* with the atomic levels of the

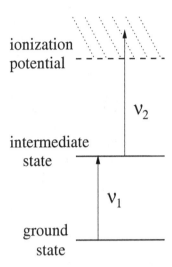

ionization
potential

ν_2

intermediate
state

ν_1

ground
state

Fig. 3.4   Schematic representation of the ion extraction method in RIMS. In this picture one photon of frequency ν_1 excites an atom to an intermediate state and a second photon of frequency ν_2 brings the atom above its ionization potential.

element. This is repeated two or three times, until the energy of the atom is above its ionization potential, electrons are freed and an ion is created. This method must be applied to material in the gas phase, thus the solid grains are first vaporised into a plume of neutral atoms and molecules by laser-induced thermal desorption. Since each element has a unique energy level structure, RIMS provides an ionization method that selects which element is going to be ionised and hence mass interferences are automatically avoided. The RIMS technique has extremely high sensitivity so that enough ions are extracted, even in the case of trace elements, to allow a relatively precise measure of the isotopic composition of the element.

CHARISMA has been applied to date to the measurement of Zr [203], Mo [204], Sr [205], Ba [245] and Ru [246] in single presolar large SiC grains, and of Zr and Mo in graphite grains [206] from the Murchison meteorite. These data are of invaluable significance in the study of the nucleosynthetic processes that produce elements heavier than Fe, as will be discussed in detail in Chapter 5 and Secs. 6.2 and 4.8.3. A similar technique has also been used to characterise polycyclic aromatic hydrocarbons (PAH) molecules in graphite grains [187].

3.4 Location and analysis of rare types of presolar grains

Presolar oxide grains are difficult to locate because they have to be recognised among the large number of oxide grains of solar-system origin. The first meteoritic oxide grains were located by analysing a meteoritic residue dispersed on a surface with a scanning electron microscope, searching for those grains among many thousands that would correspond to the required type. The selected grains were subsequently analysed with the ion microprobe in order to establish their isotopic composition and hence their possible stellar origin [136]. This procedure is extremely demanding with regards to operational time, and faster techniques have been recently implemented for the search of presolar oxides and other rare presolar grains. A fast procedure used at the Max-Planck-Institute for Chemistry in Mainz and at Washington University in St Louis is dedicated to the extraction and analysis of rare types of presolar grains. This technique was first applied successfully to a grain-rich residue from the Tieschitz meteorite by Nittler *et al.* [212] to recover a relatively large number of presolar oxide grains.

The analysis involves three different phases. First, candidate presolar oxide grains showing exotic composition are located by *ion imaging*, an operational feature of the ion microprobe that allows one to "photograph" the isotopic composition of large areas of the sample. The meteoritic residue is dispersed onto a surface resulting in a random distribution of grains, among which are those of rare types. The sample area on which the grains are deposited is then "mapped" as a function of the abundance of a given isotope. This is done by defocusing the primary beam so that it illuminates a relatively large area (e.g. 100 m in diameter). Secondary ions are emitted from different spots within this area and, by going through the electrostatic lenses, produce an electronic image which is converted to an optical image via a microchannel plate and a phosphor screen. Since the secondary ions go through the mass spectrometer, they are mass separated and in this way it is possible to obtain the image of a selected isotope. A magnified image at high spatial resolution of the sample surface is thus produced, showing the regions of the area where the given isotope is more abundant. These regions represent grains composed by the given mapped element.

For example, from an image mapping the abundance of ^{28}Si, such as the top panel of Fig. 3.5, it is possible to locate the position of SiC grains, while for an equivalent image mapping the abundance of ^{16}O it is possible to locate the position of oxide grains. Isotopically anomalous oxide grains are identified among all oxide grains by mapping the abundance of ^{18}O, as

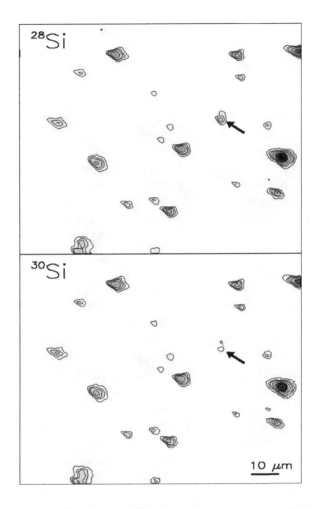

Fig. 3.5   Ion images of ^{28}Si (top) and ^{30}Si (bottom) covering an area of $100 \times 100 \ \mu m^2$. The grey colour scales from darker zones, where the selected isotope abundance is higher, to lighter regions where the abundance is lower, with overlayed contours. The colour is scaled so that grains with isotopic abundances close to solar look similar in both images. A SiC-X grain, characterised by depletion in ^{30}Si with respect to solar, is indicated by the arrow (courtesy Larry Nittler).

well as that of ^{16}O. Sixty oxide grains were found by Nittler *et al.* [212] as candidates to be attributed a presolar origin because their ^{16}O/^{18}O ratio deviated by more than 3σ from the average of the Gaussian distribution fitted to all the 17,000 mapped grains. Of these, 38 grains were confirmed to have anomalous composition, indicating a presolar stellar origin, by detailed analyses of all three O isotopes. Some of the remaining presolar candidates were actually destroyed during the search, which represents a drawback of the ion mapping technique.

Another problem is the fact that presolar grains with $^{16}O/^{18}O$ ratio close to normal cannot be identified, and some presolar grains could have a $^{16}O/^{18}O$ ratio close to solar, but an anomalous $^{16}O/^{17}O$ ratio. Performing the ion mapping of the ^{17}O abundance is infeasible because its abundance is very low and because mapping is performed at a mass resolution too low to be able to resolve ions of ^{16}OH molecules from ions of ^{17}O atoms. To evaluate the fraction of presolar candidate oxide grains lost to further analysis by the mapping technique because of not having anomalous $^{16}O/^{18}O$, Nittler *et al.* [212] measured $^{16}O/^{17}O$ ratios in a different sample by high mass resolution mapping. To this aim a "semi-automated" method was developed by which first the coordinates of each oxide grain where determined and then analysis was done at high mass resolution only at such coordinates and for a short time. In this way it was possible to estimate that $1/4$ to $1/2$ of presolar grains were missed because of not having anomalous $^{16}O/^{18}O$.

Images of the presolar candidate grains could be obtained by scanning electron microscope (SEM). In this phase it was also possible to determine the chemical composition of the grains using X-ray spectroscopy. All the grains analysed were found to be corundum grains (Al_2O_3), except for one grain which also had a Mg X-ray peak, on top of those of O and Al, and was hence recognised as spinel ($MgAl_2O_4$). As a final step, after the analysis with SEM, each grain was analysed individually in the ion microprobe for its oxygen isotopic ratios and the initial ratio of the unstable ^{26}Al to the stable ^{27}Al, as inferred from the excess of ^{26}Mg, the decay daughter of ^{26}Al.

The ion imaging technique described above has also been successfully applied to the search of SiC grains with isotopic composition different from that shown by the large majority (>90%) of SiC grains. These rare SiC grains are classified into different populations and will be discussed in Sec. 4.8. For example by mapping the abundance of ^{30}Si, as shown in the bottom panel of Fig. 3.5, it is possible to locate the position of SiC grains belonging to the rare SiC-X population ($\simeq 1\%$ of SiC grains), which are characterised by a large deficit of ^{30}Si with respect to the solar system. In the search for grains belonging to the SiC-X population a few of the candidate grains selected by ion imaging turned out to be silicon nitride grains (Si_3N_4) when analysed with the SEM. These have similar isotopic compositions to those of SiC-X grains, pointing to a SNII origin [215].

The first discovery of presolar silicate grains within Interplanetary Dust

Particles and in the Acfer 094 meteorite, which has been referred to in Sec. 1.4, was accomplished by performing ion imaging with the unprecedented sensitivity of the NanoSIMS instrument [188, 200]. Note that, in contrast to the direct ion imaging in the old instruments, where the primary ion beam is defocused, ion images in the NanoSIMS are obtained by rastering the primary ion beam over the sample.

A different technique for the search of presolar grains of rare type has been developed at the California Institute of Technology in Pasadena (USA) by implementing an "automated machine vision system" on SEM [114]. The surface to be analysed is divided into small areas of about 100×100 μm^2, which are scanned one by one by the electron beam. Grain locations are identified as regions of higher intensity of secondary electrons, and recorded. The electron beam is redirected to the highest intensity point of each of these regions to obtain information on the grain chemical type through the X-ray spectrum. The sample, and the coordinates of the grains of interest, is then moved to the ion microprobe for isotopic analysis. Presolar grains are finally recognised on the basis of their isotopic composition. Fourteen presolar corundum grains from the Bishunpur and Semarkona meteorites were located with this technique [58] as well as two presolar hibonite grains ($CaAl_{12}O_{19}$) [59].

Recently, a system of fully automated isotopic measurements at high-mass resolution has been implemented on the ion microprobe at the Canergie Institution of Washington (USA) [210]. First isotopic images are produced by scanning a given small area of about 100×100 μm^2. Then, the coordinates of individual particles are located. The primary ion beam is then focused on each particle, which is hence analysed for its isotopic composition. The process is then repeated for another small area. This system allows collection of a large amount of data in a relatively easy way and it will be of much use in expanding the current set of data, and subsequently in finding more of the rare types of presolar grains.

3.5   Concluding remarks

The ideal information from presolar grains is represented by a complete dataset of the isotopic composition of all the elements present in each single grain. As the sensitivity of the instruments used to perform analysis of presolar grains increases, it is likely that this goal will be achieved in the near future. Each of these datasets will have to be explained consistently by the model of a given stellar environment. They will represent very stringent and precise constraints on nucleosynthesis and stellar evolution theories.

3.6 Exercises

(1) What mass resolution is needed to separate the ions of ^{17}O atoms from the ions of ^{16}OH molecules? (The mass excess Δ of ^{16}O and ^{17}O are -4.737 MeV and -0.809 MeV, and that of H is 7.289 MeV)

(2) If using an electric field to separate masses, how long is the time taken by the ^{17}O ions to cover a given distance in the electric field, with respect to the time taken by the ^{16}OH ions? (Suppose that both ^{16}OH and ^{17}O ions have charge equal one.)

If using a magnetic field to separate masses, how large is the radius of the orbit taken by the ^{17}O ions in the magnetic field, with respect to the radius taken by the ^{16}OH ions? (Suppose that the ions are accelerated to a given kinetic energy by an electric field of potential V before entering the magnetic field.)

(3) Calculate the speed of a SIMS primary ion beam made of Cs+ ions ($A = 133$) accelerated to an energy of 20 keV.

Chapter 4

The Origin of Presolar SiC Grains

In this chapter we look into the details of the composition of presolar SiC grains. First, I discuss the classification of SiC grains in several groups on the basis of their C, N and Si compositions. Then, the origin of the majority of SiC grains is investigated. Using some of the basic nucleosynthesis tools from Chapter 2, we will see how the compositions of SiC grains can be used as valuable constraints on theoretical models, in particular of the evolution and nucleosynthesis of stars of low mass.

4.1 Classification of SiC grains on the basis of their C, N and Si compositions

Thousands of single SiC grains of relatively large size, from 1 to 5 μm, i.e. from the KJE to the KJH samples in Fig. 3.1, have been analysed for their carbon, nitrogen and silicon compositions. The isotopic distributions for these elements are shown in Figs. 4.1 and 4.2. Recent NanoSIMS measurements have shown that SiC grains of size smaller than those shown in the plots have very similar composition distributions [11]. The data reported in the plots have been collected from grains recovered from the Murchison meteorite. However, presolar SiC grains recovered from Orgueil, Indarch and other chondritic meteorites do not present any intrinsic difference in their composition from the SiC grains from Murchison [2, 135, 240], when taking into account the differences in the distributions of the grain sizes. For unknown reasons, Murchison grains are, on average, much larger than those in the other meteorites.

The Si isotopic ratios in Fig. 4.2 are reported, as usual, in terms of parts per thousands, or *permil* ($^o/_{oo}$), variation with respect to solar. For example, for the $^{29}Si/^{28}Si$ ratio:

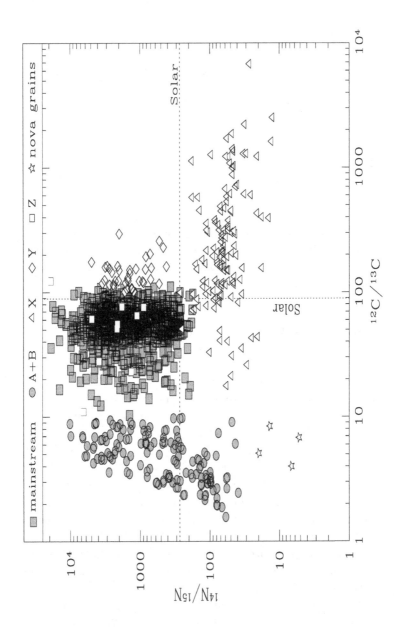

Fig. 4.1   Carbon and nitrogen isotopic ratios measured in single SiC grains (size > 1μm). The different populations covering different regions of the plot, as discussed in the text, are represented. The solar composition is shown by the dotted lines. (Data tables courtesy Sachiko Amari and Larry Nittler).

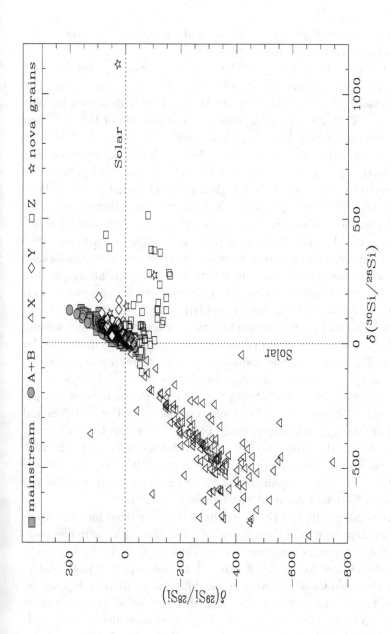

Fig. 4.2 Silicon isotopic ratios measured in single SiC grains (size > 1μm). The different populations covering different regions of the plot, as discussed in the text, are represented. The solar composition is shown by the dotted lines. The δ notation is explained in the text. (Data tables courtesy Sachiko Amari and Larry Nittler).

$$\delta(^{29}\text{Si}/^{28}\text{Si}) = \left(\frac{(^{29}\text{Si}/^{28}\text{Si})_{measured}}{(^{29}\text{Si}/^{28}\text{Si})_{solar}} - 1 \right) \times 1000,$$

where the abundant ^{28}Si, which represents about 92% of all silicon in the solar system, is used as the reference isotope.

Since the astrophysical origin of the grains imprints a signature on their isotopic composition, a first step in order to identify where SiC grains came from is to classify them into different subgroups on the basis of their isotopic composition. This has been done using the composition of the two main elements composing SiC, i.e. silicon and carbon, and of a third element present in relatively large abundance, in this case nitrogen. By combining the information presented in Figs. 4.1 and 4.2, presolar SiC grains have been observed to cluster in different regions of the plots and thus have been divided into several subgroups, or *populations*. It is important to remember that such classification is meant to be a tool for further investigation and thus it is not to be considered as a rigid, unchangeable structure. For example, early on, some grains were placed into two distinct populations: A and B, while now it has been realised that these two populations produce a continuum region in their carbon and nitrogen isotopic composition, and hence are now considered together as population A+B. Future insights may provide hints for making further modifications to the present classification.

The largest of the SiC grain populations comprises more than 90% (by number) of SiC grains, and the grains belonging to it are thus named *mainstream* grains. These grains are identified by their ^{12}C/^{13}C ratios in the range 10 to 100, the solar ratio being 89, and their silicon isotopic composition within -5% and $+20\%$ of the solar composition. The ^{29}Si/^{28}Si and ^{30}Si/^{28}Si are strongly correlated, with the ^{29}Si/^{28}Si ratios typically being about 30% higher than the ^{30}Si/^{28}Si ratios, with respect to solar [122, 127], the data points thus lie along a line of slope $\simeq 1.3$ (the *mainstream line*). The ^{14}N/^{15}N ratios of mainstream SiC grains range between 200 and 20,000, the terrestrial ratio used as reference being 272.

The remaining $\simeq 10\%$ of the grains have been classified into five other populations, represented in Figs. 4.1 and 4.2 together with the mainstream grains. Their isotopic characteristics are illustrated in Table 4.1. Note that the abundances for grains of size > 1 μm are reported in the table. However, the abundance of grains in the different population depends on grain size, for example among grains of smaller size, SiC-Z grains are more abundant, up to 3% [127]. The A+B grains are mainly identified by their low ^{12}C/^{13}C ratios [15]. The X grains show a wide range of ^{12}C/^{13}C ratios,

Table 4.1   Isotopic characteristics of the populations of presolar SiC grains.

Name	Abundance	$^{12}C/^{13}C$	$^{14}N/^{15}N$	$\delta(^{29,30}Si/^{28}Si)$
mainstream	>90%	10 – 100	200 – 20,000	−50 to +200, $\delta(^{29}Si/^{28}Si) \simeq$ 1.3 $\delta(^{30}Si/^{28}Si)$
A+B	∼5%	<10 (B) <5 (A)	39 – 10,000	∼ mainstream
X	≃1%	10 – 10,000	10 – 200	negative, down to −700
Y	≃1%	>100	∼ mainstream	typically $\delta(^{29}Si/^{28}Si)$ $< \delta(^{30}Si/^{28}Si)$
Z	≃1%	∼ mainstream	∼ mainstream	typically $\delta(^{29}Si/^{28}Si)$ $<< \delta(^{30}Si/^{28}Si)$
nova	4 grains	4 – 10	5 – 20	$\delta(^{29}Si/^{28}Si)$ $< \delta(^{30}Si/^{28}Si)$

low $^{14}N/^{15}N$ ratios and strong enhancements in ^{28}Si [8]. The Y grains are defined by having $^{12}C/^{13}C$ ratios greater than 100, and the fact that they typically lie to the right of the mainstream correlation line [14]. The Z grains are characterised by $^{30}Si/^{28}Si$ ratios typically much higher than their $^{29}Si/^{28}Si$ ratios, with respect to solar [124]. Four remaining grains could not be included in any of the populations described above and have been classified as possible *nova grains*. They are, in fact, enriched in the typical products of nova nucleosynthesis: ^{13}C, ^{15}N and ^{30}Si [7]. A very unusual grain was also found, which is not included in Fig. 4.2, showing extremely high $^{29}Si/^{28}Si$ and $^{30}Si/^{28}Si$ ratios corresponding to δ-values $\simeq$ 3000$^o/_{oo}$[17].

The origin of mainstream SiC grains and the composition of elements from He to Ti in mainstream SiC grains are discussed in the main part of this chapter. The composition of heavy s-process elements in mainstream SiC grains represents the topic of the next chapter. The origin of the other smaller populations of SiC grains will be briefly addressed in the final section of this chapter (Sec. 4.8).

4.2   Where did mainstream presolar SiC grains come from?

Two main clues point to the astrophysical origin of the majority of presolar stellar SiC grains. The first clue is that, typically, SiC forms in a carbon-rich gas, i.e. in which the carbon/oxygen ratio is equal to or greater than unity. We have already noticed (see Sec. 1.4) that C/O $\geq$ 1 is a special condition, not easily reproduced in astrophysical sites. This is because the major producers of C and O in the Galaxy, massive stars exploding as supernovææ of type II, always produce more oxygen than carbon through complete He burning. Hence, the C/O ratio in the Galaxy is usually lower than unity, e.g. it is 0.4 in the Sun. The second clue for the origin of SiC grains is that they are the carriers of the anomalous Xe-S component (see Sec. 1.2), pointing to a formation site where the s process occurs. Since mainstream SiC grains represent the vast majority of SiC grains, they are statistically most likely to be the carriers of the Xe-S anomalies. Early measurements of SiC grains in bulk, of the isotopic composition of several heavy elements present in trace amounts, Kr [167], Sr [219], Ba [218], Nd, Sm [234, 235, 305] and Dy [235], also showed a clear s-process signature, with the abundances of p- and r-only nuclei always being extremely low with respect to the abundances of s-only nuclei, as compared to the solar system composition.

Since the 1950s, observations showed that s-process elements, including the unstable element technetium, are enhanced in some spectral types of red giant stars, also referred to as *chemically peculiar red giants*. In the current standard spectral classification scheme, which was developed at Harvard Observatory in the early 20th century, stars are classified on the basis of their surface temperature. For example, stars (among which the Sun) showing the absorption lines of metals such as calcium have a temperature of the photosphere between 5,000 to 6,000 K, because in that range of temperature the observed lines are populated. Their colour is yellow and they are classified as G-type stars. In some giant stars the absorption lines of complex molecules such as TiO are observed. This means that the temperature of the star is low enough, less than 3,500 K, to allow the formation of such molecules. These are red in colour and classified as M-type stars. The chemically peculiar red giant stars of interest here belong to the category of M stars, in terms of their temperature. However, they have a special classification of their own due to their unusual chemistry. Stars of type M showing the presence of ZrO molecules in their atmospheres are defined as S stars. The presence of ZrO molecules implies that heavy elements such as Zr, which is typically produced by the s process, are enhanced in the atmospheres of these stars, as clearly shown by the element abundances derived spectroscopically [48, 255]. Another chemically peculiar class of stars are C stars, where C is for carbon. These stars have spectra dominated by lines of carbon compounds such as C_2, CH, CN and TiC, indicating that $C/O > 1$. In particular, C(N) stars, a type of C stars, are cool red giant stars very similar to M-type stars. Since in the atmospheres of C(N) and S stars both s-process elements and carbon are enhanced, S and C(N) stars are theoretically interpreted as the same type of stars at different points of their evolution. In particular, C(N) stars are believed to be S stars that have further evolved. Stars of type C(N) fulfil the two basic requirements to be the best candidate for the site of origin of presolar mainstream SiC grains: they are rich both in carbon and in s-process elements [1]. The existence of dust from C(N) stars in the solar system was proposed already a decade before the discovery of SiC grains, on the basis of the s-process signatures in primitive meteorites [62, 71, 260].

A large set of evidence has built up to date indicating that C(N) stars are the parent stars of the vast majority of presolar SiC grains. Evidence that SiC molecules are present in the atmospheres of C(N) stars is shown by the observation of their characteristic emission line at 11.2 μm [74, 259, 282]. Further proof is that the distribution of the $^{12}C/^{13}C$ ratio in mainstream SiC grains matches that observed in C(N) stars [160] (see also Fig. 3 of Ref. [126]), with most of the measurements occurring in the range of $^{12}C/^{13}C$ between 50 and 60 both in mainstream SiC grains and in C(N) stars. For elements lighter than Fe in general, there are qualitative agreements between the predicted composition of C(N) stars and the data from SiC grains. These are discussed in the remaining of this chapter.

4.2.1 Theoretical modelling of AGB and C(N) stars

In the theoretical framework of stellar evolution, stars of types S and C(N) are identified as stars on the Asymptotic Giant Branch (AGB), which represents a late stage of the evolution of stars of mass lower than $\simeq 8\ M_\odot$. The structure of AGB stars has been introduced in Chapter 2, Sec. 2.2, and shown in Fig. 2.4. Here, I describe in more detail the evolutionary path by which low-mass stars reach the AGB phase and the structure and properties of this type of star.

The theoretical evolution in the Hertzsprung-Russell (H-R) diagram of a star of 3 $M_\odot$ and $Z = 0.02$ ($\sim$ solar) from its birth to the AGB phase is shown in Fig. 4.3. Hydrogen exhaustion in the core marks the end of the long and stable main sequence phase in the life of a star. During the very last phases of core H burning the star goes through a phase of overall contraction, from point A to point B in Fig. 4.3 and H burning begins in a shell around the He core. When the mass of the He core is about 10% of the total stellar mass, the centre starts contracting strongly under its gravitational force and the star expands by more than 100 times its initial size. The effective temperature drops, moving the star to the red giant branch, far to the right in the H-R diagram. It is at this point that, for the the first time, mixing of material from the internal layers of the star modifies the composition of the stellar surface during the *First Dredge-Up*. While the convective envelope expands outwards, its bottom border reaches down to inner regions of the star where the temperature had been high enough, during the main sequence, to change the composition of several isotopes via proton captures (Sec. 4.3).

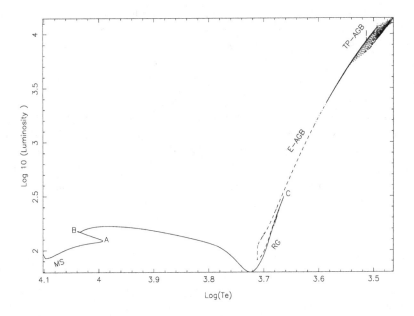

Fig. 4.3   Hertzsprung-Russell diagram, in which the stellar luminosity (on a Log_{10} scale) is plotted as a function of the effective (i.e. surface) temperature (also on a Log_{10} scale, and decreasing to the right), showing the theoretical evolution of a star of $Z = 0.02$ and mass 3 $M_\odot$. The label MS is for main sequence, RG for red giant, E-AGB for early Asymptotic Giant Branch and TP-AGB for thermally pulsing Asymptotic Giant Branch (see text for details and explanation of points labelled A, B and C).

At the point labelled C in Fig. 4.3, He starts burning in the centre. Because the temperature increases very fast from values at which He does not burn, below $\simeq 1 \times 10^8$ K, to $\simeq 2 \times 10^8$ K, where He burning is very efficient, and because the energy production for the 3α reaction has a strong dependence on the temperature ($\sim T^{40}$), the huge and sudden energy release drives a central convective zone, which almost reaches the location of the H-burning shell. A central source of energy tends to move the star back toward the main sequence region, with higher effective temperatures, while the presence of two energy sources (core He burning and H-shell burning) bring the star toward lower luminosities. When He is also exhausted in the centre, He-burning starts in a shell around the core, now degenerate and rich in C and O produced by complete He burning. The structure completes the loop reaching back close to point C in the figure, and the star ascends the Asymptotic Giant Branch. This is so-named because the evolutionary track at this point approaches the first giant branch almost asymptotically. In stars of mass greater than $\simeq 5\ M_\odot$, after the exhaustion of He in the centre, the convective envelope penetrates inwards again mixing internal material to the surface during the *Second Dredge-Up*.

The AGB evolution is divided into two phases: the early-AGB (E-AGB) and the thermally pulsing AGB (TP-AGB, which we will refer to hereafter simply as AGB). The onset of the thermal pulses described below is the dividing time between these phases. The evolution in time and in mass of the points defining the structure of AGB stars (see Fig. 2.4) is presented in Fig. 4.4. In calculation of stellar evolution the variable M_r, representing the mass contained within a sphere of given radius r, is used as the independent variable, rather than the radius r. This is because this variable is independent of variations in the geometry of the structure, such as expansions or contractions, and better represents a particular element of matter within the stars. Thus, in Fig. 4.4 the evolution of the stellar structure is represented using a mass coordinate.

Since the advent of computers in the 1960s, it has been possible to perform detailed evolutionary calculations capable of representing the complex structure of Asymptotic Giant Branch stars. The structure evolves in time in a very peculiar way as the H- and the He-burning shells are activated alternately. Models have shown that during the AGB phase the H-burning shell dominates the energy production for most of the time. Hydrogen is transformed into He at the top of the He intershell, whose thickness consequently grows. The bottom layers of the intershell are compressed until their temperature and density become high enough that He burning is triggered in an almost explosive way. The thermal runaway, also known as thermal instability, or thermal *pulse*, generated by this sudden release of energy causes the energy transport in the whole He intershell to turn from radiative to convective and the material is mixed and homogenised throughout the region (*convective pulse*). In the meanwhile the H shell cools and

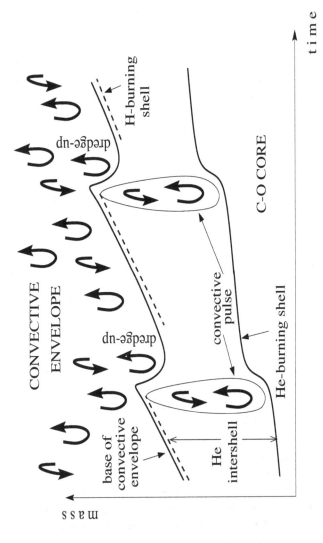

Fig. 4.4   Schematic evolution in time (x-axis) of the location in mass (y-axis) of the points defining the structure of an AGB star (see Fig. 2.4) and defined as follows. The bottom solid line represents the He-burning shell, corresponding to the upper edge of the C-O core, the dashed line represents the H-burning shell, corresponding to the upper edge of the He core (usually referred to as M_{core} or M_H), the upper solid line represents the bottom edge of the H-rich convective envelope.   The He intershell is the region, rich in He, located between the two burning shells.   Thick arrows represent convection.   Convective zones develop episodically in the He-rich region.   Arrows labelled as "dredge-up" represent episodes of mixing of the carbon-rich He intershell material into the envelope.

H burning stops. This phenomenon was first discovered by Schwarzschild & Härm in 1965 [249]. A thermal pulse quenches after a few hundred years and the H burning starts again. This cycle is repeated 10–100 times with intervals between pulses of the order of $10^3 - 10^5$ years. The total number of calculated pulses depends on the initial stellar mass and on the choice of the mass-loss law (see extensive review in Ref. [142]). Recent AGB models have been calculated, among others, by Boothroyd and Sackmann [35, 36, 37, 38], Lattanzio [162, 163, 164], Blöcker [34], Forestini and Charbonnel [92], Straniero *et al.* [265], Mowlavi [197], Herwig [115], Stancliffe *et al.* [262] and Ventura and D'Antona [284]. The most recent review on AGB models has been presented by Herwig [116]. A book focused on AGB stars, covering all their aspects and their astronomical implications has been edited by Habing & Olofsson [109].

Just after the quenching of a thermal pulse an episode of mixing may occur, known as *Third Dredge-Up* (TDU), during which a fraction of the material from the He intershell is carried to the envelope (see Fig. 4.4). During the thermal pulse partial He burning produces high amounts of carbon in the He intershell. This carbon is then mixed into the envelope by TDU, allowing AGB stars to become carbon-rich after a certain number of TDU episodes. However, there are some constraints to the allowed masses of carbon stars. At solar metallicity, it is predicted that TDU is not efficient enough to produce a carbon star when the initial mass of the star is lower than about 1.5 $M_\odot$ [164, 265]. However, this lower limit decreases for models of lower metallicity as the TDU become more efficient and it is easier to alter the initial composition of the star. For stars with initial masses above $\simeq 5$ $M_\odot$, models have shown that a phenomenon known as *hot bottom burning* prevents the formation of a carbon-rich envelope [40]. In these intermediate-mass stars the base of the convective envelope reaches temperatures of up to 80 million degrees. Under these conditions the CN cycle is activated, carbon is converted into nitrogen by proton captures and the C/O ratio remains below unity. At solar metallicity, the mass range at which carbon stars are expected to form is from $\simeq 1.5$ to $\simeq 4 - 5$ $M_\odot$ [106].

Also of much importance, during the AGB phase the star experiences strong stellar winds. Observed mass-loss rates are from 10^{-8} $M_\odot$/yr up to 10^{-4} $M_\odot$/yr during the advanced AGB phases, as compared to mass loss from the Sun of about 10^{-14} $M_\odot$/yr. Many AGB stars are observed to be pulsating variable stars belonging to the Mira class, showing periodic variations in luminosity and radius with periods of the order of 500 days. Outward shock-waves are produced in their atmospheres so that material is periodically hit, and shells of gas and dust are created. These phenomena, together with the presence of dust, are believed to be responsible for the stellar winds (see e.g. [45]). The dust-rich circumstellar shells generated by the expelled material are observed around carbon stars. When the dust shells become so thick that all the visual light from the star is absorbed and re-emitted in the infrared, the star is obscured in the optical wavelengths. After the stellar envelope is completely expelled by fast stellar winds, a planetary nebula forms, which is characterised by a shell of material surrounding the C-O degenerate core of the former star, which remains as a cooling white dwarf.

4.3   Carbon and nitrogen in mainstream SiC grains and in AGB stars

Because the isotopic compositions of carbon and nitrogen are easily modified in stellar interiors by proton capture via the CNO cycle, it is not surprising that mainstream SiC grains show such a wide range of variation in the isotopic composition of these elements.

   The first time that the CNO isotopic ratios are modified at the stellar surface is when the first dredge-up occurs during the red giant phase, carrying material from the deep layers to the surface of the star (see e.g. Refs. [27, 39, 79, 89]). During the main sequence phase, proton captures occur in the deep layers of the star, where the temperature is high enough to activate such reactions. The CNO composition of a stellar model of initial mass 3 $M_\odot$ and initial solar composition, as a function of the depth in mass, just before the occurrence of the first dredge-up is shown in Fig. 4.5. The CNO abundances change at different locations in mass, depending on the temperature achieved in each mass layer during the previous evolution. Moving deeper into the star, the temperature increases and proton captures become efficient. First, ^{12}C is destroyed, while a "bump" is produced in the abundance of ^{13}C, at $\simeq$ 1.4 $M_\odot$ in Fig. 4.5, and the abundance of ^{15}N is reduced by more than an order of magnitude. Deeper in the star ^{13}C is converted into ^{14}N. At much deeper locations, below the H-burning shell located at $\simeq$ 0.3 $M_\odot$ in Fig. 4.5, CNO isotopes are completely processed by the CNO cycle and have thus reached the equilibrium values, resulting in the conversion of C and O into N. As for the O isotopes, the abundance of ^{16}O is mostly unchanged down to the location of the H-burning shell, the ^{17}O abundance presents a bump at $\simeq$ 0.5 $M_\odot$, while ^{18}O is destroyed

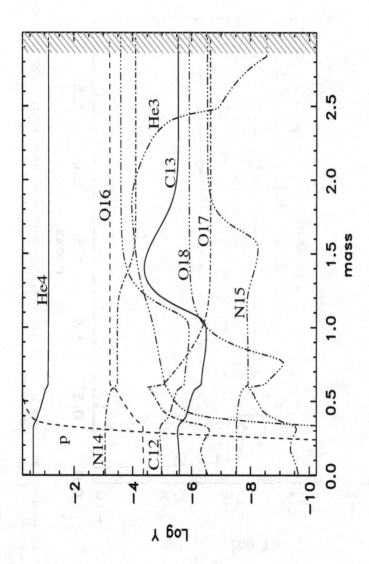

Fig. 4.5 Abundances of the CNO isotopes on a logarithmic scale as a function of the position in mass within the star just before the occurrence of first dredge-up (stellar model of 3 $M_\odot$ and solar initial composition). The shaded region represents the convective envelope. (Note that in this calculation the initial abundance of ^{3}He was set as zero, rather than the solar value of $\simeq 10^{-5}$).

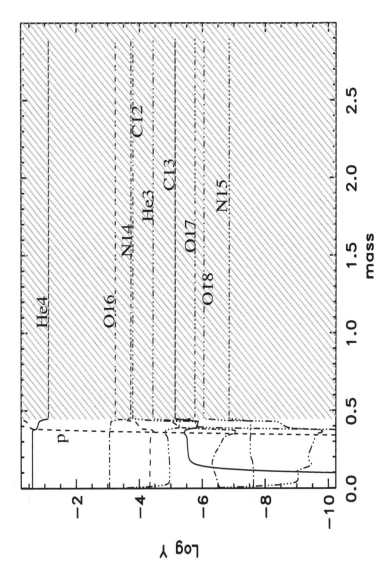

Fig. 4.6 Abundances of the CNO isotopes on a logarithmic scale as a function of position in mass within the star at the time when the convective envelope (shaded region) reaches its deepest penetration in mass during the first dredge-up (stellar model of 3 $M_\odot$ and solar initial composition).

below $\simeq 1\ M_\odot$.

During the first dredge-up, in the stellar model of 3 $M_\odot$ shown in Fig. 4.6, the convective envelope reaches down to a mass of 0.5 $M_\odot$ and thus all the abundances from the surface down to 0.5 $M_\odot$ are homogenised by convection. The CNO composition at the surface of the star is modified, resulting in the variations in the surface abundances ratios presented in Table 4.2.

Table 4.2 CNO isotopic changes after first dredge-up at the surface of a 3 $M_\odot$ star.

isotopic ratio	solar	after first dredge-up
$^{12}C/^{13}C$	89	25
$^{14}N/^{15}N$	272	1400
$^{16}O/^{17}O$	2660	340
$^{16}O/^{18}O$	500	650

In Fig. 4.7 the evolution of the carbon and nitrogen isotopic ratios predicted at the stellar surface during the life of low-mass stars is compared to the compositions measured in mainstream SiC grains. The effect of the first dredge-up in the 3 $M_\odot$ stellar model is represented by the solid line changing the initial solar composition to $^{12}C/^{13}C \simeq 25$ and $^{14}N/^{15}N \simeq 1400$ (see Table 4.2). In the 1.5 $M_\odot$ model, the envelope reaches less deeply into the internal stellar layers and hence the predicted $^{14}N/^{15}N$ remains closer to solar, at $\simeq 900$. Consequently, also the $^{12}C/^{13}C$ ratio should be closer to solar for this stellar model. However, spectroscopic observations have shown that, instead, the $^{12}C/^{13}C$ ratio in red giant stars after the first dredge-up decreases with the stellar mass, down to values lower than 10 [102]. Thus, in Fig. 4.7 the predictions for the 1.5 $M_\odot$ stellar model are forced to $^{12}C/^{13}C = 12$ at the beginning of the AGB phase.

The low $^{12}C/^{13}C$ ratios that are observed in red giant stars of low masses, cannot be explained in the framework of the first dredge-up process, which instead predicts that the $^{12}C/^{13}C$ ratio should increase for stars of lower masses. It has thus been proposed that some "extra-mixing" processes occur in red giant stars, after the first dredge up, by which material from the base of the convective envelope penetrates the underlying radiative region, which is by definition stable against convection. There, the temperature is high enough to further process some ^{12}C into ^{13}C [39, 55, 83, 269]. In this way it is possible to lower the $^{12}C/^{13}C$ ratio to the observed values. This process can only occur in stars of mass lower than approximately

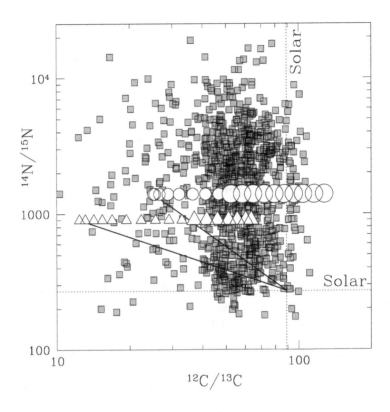

Fig. 4.7   The C and N compositions measured in single mainstream SiC grains (squares, from Fig. 4.1) are compared to the evolution of the isotopic ratios predicted at the stellar surface for stellar models of 3 $M_\odot$ (circles) and 1.5 $M_\odot$ (triangles) with initial solar composition during the red giant and AGB phase. The solid lines represent the change in composition starting from solar initial values to the values produced by the operation of the first dredge-up and extra-mixing phenomena during the red giant phase (see text for details). The open symbols represent the change in composition during the AGB phase because of the operation of the third dredge-up. The larger symbols are employed when the condition for the formation of SiC, C/O > 1, is satisfied in the envelope.

2.3 $M_\odot$. This is because, in this case, the red giant phase lasts long enough to allow the H-burning shell to progress to a point in mass where it can destroy the barrier to mixing due to the composition discontinuity left behind by the first dredge-up (visible at around 0.5 $M_\odot$ in Fig. 4.6). This type of non-standard mixing could be generated by rotation and/or magnetic fields. However, exactly what drives the extra mixing, and how it works is still largely unknown.

When the star ascends the Asymptotic Giant Branch, the third dredge-up carries ^{12}C to the surface (Sec. 4.2.1). The star can become carbon rich and hence SiC grains can form. By the end of the AGB phase, the ^{12}C/^{13}C ratio reaches values higher than 100 in the case of the 3 $M_\odot$ star, and between $\simeq$ 40 and 60 in the case of the 1.5 $M_\odot$ star. The N and O ratios are unaffected by the third dredge-up.

As shown in Fig. 4.7, the composition measured in SiC grains can be qualitatively explained using the theoretical predictions of nucleosynthesis and mixing occurring during the red giant and AGB phase described above. This is particularly true if some extra mixing during the red giant phase is included in the models, as required by the stellar observations discussed above. However, the range of isotopic ratios resulting from the models, with the condition that C/O > 1, is far too small to explain the whole range covered by SiC grain data. The measured ^{12}C/^{13}C ratios spread to lower values than those predicted, while the ^{14}N/^{15}N ratios spread both above and below the theoretical lines.

The extra-mixing phenomena described above could in principle be at work not only during the red giant phase, but also during the AGB phase (in this case they are sometimes known as *cool bottom processing*). Actually, they seem to be required to match the O isotopic composition of a fraction of presolar oxide grains [289] (Sec. 6.3). This further CNO processing can also explain SiC grain data with lower ^{12}C/^{13}C ratios and higher ^{14}N/^{15}N ratios than those covered by the points predicted by the models in Fig. 4.7 [216]. Note that there are no limits for the stellar masses at which extra mixing could occur during the AGB phase, due to the time-scale associated with the advancement of the H-burning shell. In these stars, for any initial mass, the H-burning shell has already advanced enough in mass to have destroyed the discontinuity in composition left behind by the first dredge-up.

On the other hand, grains showing ^{14}N/^{15}N ratios lower than 1000 are difficult to account for, since production of ^{15}N in AGB stars is not predicted to occur. They could be explained by variations in the initial composition of the star. In fact, the ^{14}N/^{15}N ratio is expected to change during the life of the Galaxy. However, the Galactic chemical evolution of the N isotopes is very complex and still uncertain (see e.g. Ref. [238]), because the abundance of ^{14}N appears to have both a primary and a secondary component, while ^{15}N is produced in nova outburst and perhaps also in SNII. Even the solar isotopic composition of N is still uncertain within a factor of two [221].

In summary, the C and N composition of mainstream grains can be used to constrain mixing processes in red giants and AGB stars and also, possibly, Galactic chemical evolution models. Unfortunately, it has not been possible to precisely measure oxygen isotopic ratios in SiC grains because of the low abundance of oxygen atoms. If this task could be achieved in the future, then these ratios will also have to be interpreted within the framework described above. Oxide grains (see Sec. 6.3) are also believed to have originated in AGB stars, but during their oxygen-rich phases. They have been used so far as complementary information to SiC grains to study the evolution of CNO isotopes in low-mass stars.

4.4   The Ne-E(H) anomalous component

Silicon carbide grains are the carriers of the Ne-E(H) anomalous component, with ^{22}Ne enriched by a factor of 100 with respect to solar (Sec. 1.2). Only a small fraction, $\simeq 5\%$, of SiC grains actually seem to contain noble gases, which makes it extremely difficult to measure their composition in single grains. This is explained by the fact that noble gases are extremely volatile, and hence could not have condensed in SiC grains. Instead, it is believed that these elements have been implanted into the grains, after having been ionised [168] (see Sec. 5.3).

The Ne-E(H) component was initially attributed, at least partly, to the early presence of the radioactive nucleus ^{22}Na, which decays into ^{22}Ne with a half-life of 2.6 years [61]. However, once it was discovered that SiC grains are the carrier of Ne-E(H), and that the abundance of ^{22}Ne correlates with that of ^4He [167, 168], it became clear that the Ne-E(H) component is to be interpreted as the product of nucleosynthesis in AGB stars. In the He intershell of AGB stars, all the initial CNO abundances, which were converted into ^{14}N during H burning, are eventually turned into ^{22}Ne during He burning. As mentioned in Sec. 2.2 α-capture reactions are activated on ^{14}N and ^{18}O at the temperature of He burning producing ^{22}Ne, and AGB stars represent a major production site of ^{22}Ne in the Galaxy.

Figure 4.8 presents a *three-isotope plot* in which two isotopic ratios, with a common reference isotope, are plotted against one another. The $^4He/^{22}Ne$ ratios measured in SiC grains of different sizes in bulk are plotted as a function of the $^{20}Ne/^{22}Ne$ ratios. Note that, since measurements in bulk are performed on millions of grains, they can be only used to derive the average properties of the parent stars of the grains. The data

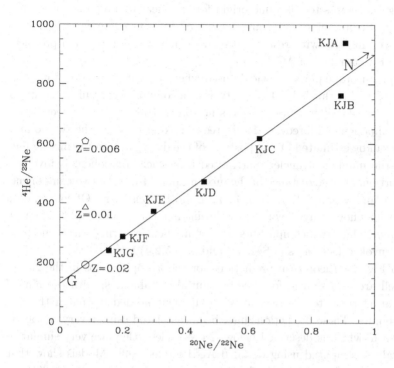

Fig. 4.8   The ^{4}He/^{22}Ne and ^{20}Ne/^{22}Ne ratios measured in SiC grains of different sizes in bulk are represented by the black squares (refer to Fig. 3.1 to relate the labels to the grain size). The solid line represents the data correlation line, which is interpreted as the mixing line between two components. To the upper right end of the mixing line lies the normal "N" component, close to the solar ratios: ^{4}He/^{22}Ne=2114 and ^{20}Ne/^{22}Ne=12.4 (out of scale in the plot). To the bottom left end of the mixing line lies the "G" component produced in the He intershell of AGB stars. The observed compositions are interpreted as having been produced by means of various degrees of mixing of the N and the G components. Theoretical predictions for the composition of the He intershell of AGB stars (G component) are shown as open symbols, and the metallicity of the star is indicated.

points lie on a straight line in the three-isotope plot of Fig. 4.8. This is interpreted as showing that the composition represented by each point was produced by a mixture between the material initially present in the envelope of the star, and the material mixed from the He intershell into the envelope by third dredge-up. These two "ingredients" are called the "N" and "G" *components*, respectively. The N component is typically taken to have a composition close to solar. The good correlation shown by the data in Fig. 4.8 points to the same origin for the measured anomalous compositions. Such compositions are dominated by a G component extremely enhanced in ^{22}Ne with respect to solar, as it is the composition produced in the He intershell of AGB stars.

Three-isotope plots are widely used when comparing isotopic ratios because of having this helpful property, that a composition resulting from the mixing of two different components lies on a straight line connecting the two components. Moreover, the degree of mixing between the two components can be estimated by the position of the data point on the mixing line, since the number of nuclei contributed from each component is inversely proportional to the distance of the mixture point from the two components (for an application of this, see Sec. 5.2.1 and Exercise 5.2). Of course, data points may not always lie on a straight line in a three-isotope plot, in which case more than two components or a component evolving with time have to be invoked (see cases in Sec. 4.6 and Sec. 5.2.4).

In Fig. 4.8 theoretical predictions for the isotopic ratios in the He intershell are also shown, for different initial metallicities, with the aim of comparing them to the average G component needed to match the SiC data points. These predictions have been calculated using a very simplistic approach, which is described below. Nevertheless, they are very similar to the values computed using detailed AGB models [99]. Models show that during partial He burning in AGB stars, He is depleted by about 30%, so that in the He intershell the typical value of the ^{4}He mass fraction is 0.7. The initial CNO abundances are taken to be equal to the metallicity of the star, and the initial abundance of ^{20}Ne also scales with the metallicity. Then, all the initial CNO abundances are converted into ^{22}Ne by H and He burning, while the abundance of ^{20}Ne is unaffected during the stellar evolution. Because both ^{20}Ne and the final ^{22}Ne scale with the metallicity, the calculated ^{20}Ne/^{22}Ne ratio in the He intershell is the same for any metallicity and it is rightly placed at the lower end of the composition observed in the grains. However, because the He-intershell ^{4}He abundance is constant and independent of the metallicity, the ^{4}He/^{22}Ne ratio increases with metallicity. It follows that stars of metallicity close to solar are the favoured candidates to be the SiC grains parent stars. This is in agreement with the fact that the carbon stars observed in the Galactic disk have metallicities within a factor of two of the solar value [160].

However, some words of caution are required: the three-isotope plot of Fig. 4.8 involves isotopes of two different elements, He and Ne. Hence it is possible that different proportions of He and Ne have actually been trapped in the grains due to some chemical fractionation effect. In this case the data would not represent the original stellar composition. Fractionation effects are measurable in the Sun where it is observed that the He/Ne in the solar wind is 0.67 of the value observed in chondritic meteorites. Applying this factor one finds that a star with initial metallicity slightly lower than solar, $Z = 0.013$, would provide a better match with the observations. However, fractionation effects are not necessarily the same in AGB stars as in the Sun. More recent work indicates that He and Ne in SiC grains do represent an implanted unfractionated component [286], and hence there would be no need to take fractionation effects into account.

The analysis of Ne in SiC grains is of further interest because of the possibility of determining the age of SiC grains from the ^{21}Ne abundance. This isotope can be produced by spallation reactions when the grains are bombarded by cosmic rays, i.e. high-energy particles such as protons, α-particles and heavier nuclei, during their passage through the interstellar medium. Excesses of ^{21}Ne with respect to the values predicted by AGB models can be related to the length of time that the grains were exposed to the cosmic rays: in other words ,the interval between their formation in the AGB stellar winds and their arrival in the solar nebula. Using this method, Lewis *et al.* [168] calculated times that SiC grains were exposed to cosmic rays (*exposure ages*) of the order of ten to a hundred million years. However, Ott & Begemann [220] have shown experimentally that the majority of presolar SiC grains would have lost essentially all the ^{21}Ne produced during spallation by recoil. These authors suggest that the observed variations of the ^{21}Ne/^{22}Ne ratios in SiC grains are more likely due to the effect of nucleosynthesis in the He burning shell of the parent AGB stars, and suggest that Xe produced by interaction with cosmic rays may be a used a future method to approach the age problem.

Lewis *et al.* [168] measured the isotopic composition of SiC grains also for the other noble gases: He, Ar, Kr and Xe. The compositions of Kr and Xe are determined by the s process and will be discussed in Sec. 5.3. The ^{3}He/^{4}He ratio measured in SiC grains is in the range 0.5×10^{-4} to 2×10^{-4}, indicating the mixing of two components: the envelope material with ^{3}He/^{4}He $\simeq 5 \times 10^{-4}$, and the He intershell material with ^{3}He/^{4}He $= 0$. The ^{3}He/^{4}He ratio in the envelope is about 10 times higher that solar because ^{3}He is produced during the main-sequence phase by the the *pp* chain in the region of the star where temperatures are between 6 and 8 million degrees (see Sec. 2.1 and Fig. 4.5). It is then mixed to the convective envelope during the first dredge-up (see Fig. 4.6). At higher temperatures all ^{3}He is converted into ^{4}He so that there is no ^{3}He present in the He intershell.

As for argon, only the ^{38}Ar$/^{36}$Ar ratio could be determined. The abundance of ^{40}Ar is extremely low and could not be measured because of the large interference of ^{40}Ca, which dominates the abundances at atomic mass $A = 40$. The ^{38}Ar$/^{36}$Ar ratio in the He intershell is changed by neutron-capture nucleosynthesis from the solar value of 0.19 to a value of $\simeq 0.6$ in solar metallicity stars [99]. However, the absolute abundances of the Argon isotopes are modified only by approximately a factor of two, so that when they are brought to and diluted in the envelope by the third dredge-up they do not substantially affect the Ar initial composition. SiC grains show ^{38}Ar$/^{36}$Ar ratios in the range 0.19 to 0.23, in agreement with the slight modification induced by the third dredge-up.

4.5   The presence of ^{26}Al

Presolar SiC grains have also been analysed for their Mg isotopic composition. While the measured ^{25}Mg$/^{24}$Mg ratios are solar within 10%, the ^{26}Mg$/^{24}$Mg ratios show large excesses that are attributed to the early presence of ^{26}Al, which decays into ^{26}Mg with a half-life of 0.7 million years. The ^{26}Al$/^{27}$Al ratios, extrapolated to the time when this element condensed in SiC grains, have a maximum measured value around 10^{-3} in the case of mainstream grains, around 10^{-2} for grains belonging to the A and B population and up to 1 in SiC grains of type X [122].

The early presence of ^{26}Al in mainstream SiC grains is related to their AGB origin. As briefly discussed in Sec. 2.1, ^{26}Al is produced during H burning if the temperature is high enough for the MgAl cycle to set in. Models have shown that ^{26}Al is produced in the H-burning shell of AGB stars up to an abundance $X_{26} \simeq 8 \times 10^{-5}$, in mass fraction. The ashes of H burning also contain ^{13}C produced by the CNO cycle. When these ashes are engulfed in the convective thermal pulse, neutrons are released by the ^{13}C$(\alpha, n)^{16}$O reaction and ^{26}Al is easily destroyed by the neutron-capture reactions (n, p) and (n, α). Depending on the maximum temperature achieved at the base of the convective pulse the ^{22}Ne$(\alpha, n)^{25}$Mg reaction can also be activated (see details in the next chapter). Hence some ^{26}Al is destroyed in the convective pulse because of the presence of a neutron flux, and only a fraction of the ^{26}Al abundance that was produced during H burning can survive to be dredged up to the surface of the star where SiC grains form. Detailed numerical models that have taken into account these effects have shown that the $^{26}Al/^{27}Al$ ratio up to $\simeq 10^{-3}$ observed in mainstream SiC grains is reproduced by AGB star models [91, 198]. Note also that the possible detection of ^{26}Al in the nearest carbon star, CW Leo [107] may provide a further link between the origin of mainstream SiC grains and carbon stars.

4.6 The puzzle of the silicon isotopic composition of mainstream SiC grains

A magnification of the region of Fig. 4.2 with isotopic ratios within 25% of the solar composition is shown in the left panel of Fig. 4.9. This is the region of the silicon three-isotope plot where the majority of SiC grain data are located: the mainstream and the A+B grains, which together constitute more than 95% of all SiC grains. The figure shows that mainstream SiC grains have a well defined distribution, from which grains belonging to the X, Y, Z and nova populations are excluded. On the other hand, grains belonging to the A and B populations are basically indistinguishable from mainstream SiC grains with regards to their silicon composition. To clearly visualise the features of the distribution of mainstream SiC grains in the silicon three-isotope plot, the right panel of Fig. 4.9 shows only mainstream SiC grains, together with their correlation line. This line, also known as the "mainstream line" has a slope of $\simeq 1.3$ and does not pass exactly through the solar composition ($\delta = 0^o/_{oo}$), but is slightly, $16^o/_{oo}$, shifted to the right of it. The error bar on each data point is lower than $10^o/_{oo}$, much smaller than the range covered by the grains. Note that the only determination of the Si isotopic composition in a carbon star is comparable with the solar composition, within a large uncertainty of $\simeq 100^o/_{oo}$ (see e.g. [153]). A fascinating and informative representation of the Si distribution is produced when plotting the data in a three-dimensional space, using the frequency of each value as the z-axis (see Fig. 12 of Ref. [210]). This representation shows that the majority of the grains lie in the middle ($\delta \simeq 50\%$) rather than at either ends of the distribution and that other local peaks are also present.

If we want to achieve a complete understanding of the information carried by presolar grains we have to explain the reason for the well-defined distribution of the silicon isotopic ratios in mainstream SiC grains. However, this distribution represents a challenging puzzle for which a conclusive, widely-accepted solution has not yet been found.

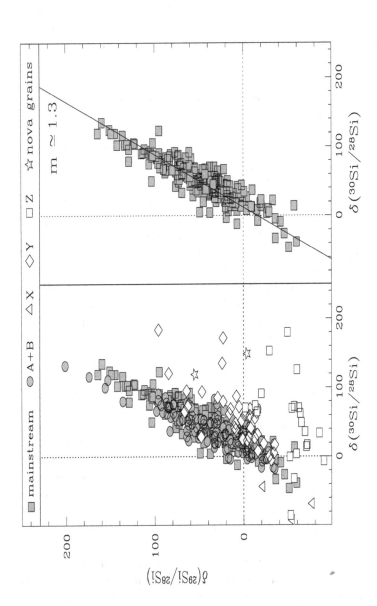

Fig. 4.9 Magnification of the region close to the solar composition of the three-isotope plot for the silicon isotopic ratios measured in single SiC grains. Since the usual δ notation is employed, the solar composition corresponds to the origin of the plot and it is shown by the dotted lines. In the left panel the data from all the SiC grain populations are showed. In the right panel only mainstream SiC grains are plotted, together with their correlation line with a slope ≃ 1.3.

To tackle the problem, first of all, it is necessary to discuss the possibility that such a distribution has been produced by nucleosynthesis occurring in the AGB parent stars of mainstream SiC grains. As in the case of the He and Ne three-isotope plot (Fig. 4.8), the mainstream correlation line could be interpreted as a mixing line along which material with an initial composition around solar is shifted toward a He intershell AGB component richer in ^{29}Si and ^{30}Si than in ^{28}Si, with respect to solar, and with the abundance of ^{29}Si 30% higher than that of ^{30}Si [264]. In this interpretation all single mainstream SiC grains could have formed in the same star, at different evolutionary times, when the anomalous He intershell material was more or less diluted with the solar envelope material.

The problem with this hypothesis is that it is not easy to explain the composition represented by the top end of the mainstream line as a product of nucleosynthesis in AGB stars. Typically, the abundances of the silicon isotopes are not much affected by nucleosynthesis in stars of low mass and metallicity around solar [177]. When neutrons are released in the convective pulse, as mentioned in Sec. 4.5, the silicon isotopes suffer neutron captures and their abundance can be modified. Such modifications are induced by the chain of reaction ^{28}Si$(n, \gamma)^{29}$Si$(n, \gamma)^{30}$Si that typically results in the production of ^{29}Si and ^{30}Si and a small depletion of ^{28}Si. There is an additional channel for the production of ^{30}Si via the reaction ^{33}S$(n, \alpha)^{30}$Si, which has a relatively high neutron-capture cross section. In the presence of neutrons, ^{33}S is produced by the ^{32}S$(n, \gamma)^{33}$S reaction starting from the abundant isotope ^{32}S. Thus in AGB stars the production of ^{30}Si is always favoured with respect to the production of ^{29}Si. This results in a mixing line with slope < 0.5, rather than $\simeq 1.3$ as for the mainstream line.

Within this interpretative scheme, a possible solution was proposed by Brown & Clayton [46]. These authors showed that in a 5.5 $M_\odot$ stellar model, if the temperature in the convective pulses is raised by 15% with respect to the standard models [140], (α, n) reactions on the Mg isotopes could produce a Si composition at the top end of the mainstream line. In this model the Ne-E(H) component is attributed to the decay of the ^{22}Na produced by the NeNa chain during hot bottom burning (see Sec. 4.2.1) at the base of the convective envelope. This scenario, however, does not account for the fact that in SiC grains the ^{22}Ne excesses are correlated with ^{4}He excesses, as discussed in Sec. 4.4, and would require much larger abundances of Na in SiC grains than observed [135]. Another major problem with this explanation is that such high temperatures would be incompatible with some of the s-process isotopic signature shown by the grains (as will be discussed in detail in Sec. 5.2.4). Moreover, the model could not account for the fact that variations in the Ti isotopic ratios have also been observed in SiC grains (Sec. 4.7), because even higher temperatures are required to affect the abundances of Ti isotopes.

The early study of Tang *et al.* [271] already pointed out the difficulty of interpreting the silicon composition of SiC grains in the framework of mixing a small number of components, and concluded that several discrete components must have been involved in creating the distribution of the silicon composition of SiC grains. If a large number of components must be invoked to explain the silicon data, then this can be achieved only by considering that not all mainstream SiC grains originated from a single star, but they must have had multiple parent stars. The variations in the silicon isotopic composition could be connected to variations in the initial composition of the parent stars, due to the effect of Galactic Chemical Evolution (GCE) on the silicon isotopes. Note that, in our quest to understand where SiC grains came from, the mixing-line interpretation discussed above, in which all the SiC grains have originated in a single star is intrinsically different from the interpretation in which SiC grains had multiple parent stars [2]. This latter interpretation is now quite generally accepted.

A possible scenario for the involvement of several stars as the SiC grain parent stars was proposed by Gallino *et al.* [101]. SiC grains with silicon isotopic ratios higher than solar were attributed to parent stars of metallicity lower than solar, down to about one-half of solar, while SiC grains with silicon isotopic ratios closer to solar were attributed to parent stars of metallicity approximately solar. The basic idea behind this interpretation was that, while ^{29}Si and ^{30}Si are only produced in massive stars which explode as SNII, only about 70% of ^{28}Si is produced by massive stars and the remaining 30% is produced during SNIa explosions, which results from the binary interaction between stars of low mass. Because massive stars evolve faster than low-mass stars, then there must be a delay in the Galactic production of ^{28}Si resulting in stars of lower metallicity having silicon isotopic ratios higher than solar.

Two years later, Timmes & Clayton [278] performed the task of quantitatively calculating the evolution of the silicon isotopes in the Galaxy on the basis of nucleosynthesis models of SNII computed by Woosley & Weaver [298] and of SNIa computed by Thielemann *et al.* [274]. A general problem related to the GCE of the silicon isotopes was found to be the fact that the solar composition is missed by the models, in that ^{29}Si is underproduced by a factor of 1.5. This factor is thus generally applied to SNII yields in order to produce consistent description of the evolution of the Si isotopes, but there has been so far no explanation for it. Note that some ^{29}Si is also missing when comparing SNII models to the composition of SiC grains of type X and graphite grains believed to have originated in such environments (see Secs. 4.8.3 and 6.2).

The GCE detailed calculations showed that the silicon evolutionary trend is the opposite of the trend proposed by Gallino *et al* [101]: the silicon isotopic ratios in the Galaxy increase with the stellar metallicity. This is due to the fact that while ^{28}Si is a *primary* nucleus, which means that it is only produced starting from the H and He initially present in the star, ^{29}Si and ^{30}Si are *secondary* nuclei, which means that their production depends on the initial metal content of the star (see Sec. 2.1.2). Hence, in spite of the fact that SNIa add ^{28}Si late in the evolution of the Galaxy, the silicon isotopic ratios are higher than in the Sun in stars with metallicity higher than solar. Since the Galaxy is enriched with metals as time progresses, this conclusion has the corollary that the parent stars of SiC grains with Si isotopic ratios higher than the Sun must have been born after the Sun. This is obviously an absurd conclusion as SiC grains were trapped in the protosolar nebula, and hence must have existed before the Sun! Another problem related to interpreting the Si compositions of SiC grains as star-to-star variations due to the effect of the GCE, is that the model predicts an evolutionary trend with slope equal to one in the silicon three isotope plot, instead of the slope equal $\simeq 1.3$ of the mainstream line [278].

To overcome these serious problems various solutions have been proposed. Clayton & Timmes [70] focused on the fact that the silicon isotopic ratios are normalised with respect to the solar ratios. This could generate the difficulty in the interpretation of Si in SiC grains, if the composition of the Sun is very different from that of the interstellar medium at the time when the Sun was born. These authors showed that if the Sun is about 25% richer in ^{30}Si than expected by the GCE, then the slope of the mainstream line could be reproduced by a GCE evolution line situated $250^o/_{oo}$ to the left of the mainstream line. The shift of the GCE line to the mainstream line should then be produced by nucleosynthesis in AGB stars, with modifications to the abundances of ^{29}Si and ^{30}Si at the level of 20% to 40%. However, it is difficult to justify the assumption that the Sun must be anomalous in its silicon isotopic ratios. Moreover, AGB models of stars of low mass and metallicity close to solar do not show the needed production of the heavy silicon isotopes.

Alexander & Nittler [3] considered the problem from a different perspective. They estimated which GCE trend of Si (and Ti) isotopes would fit the SiC grain isotopic data. They reconciled the fact that the grain parent stars appear to be younger than the Sun, by assuming that the Si composition of the Sun is atypical, having been affected by the addition or removal of a small amount of supernova material. However, their detailed analysis showed that with current SNII models it is difficult to achieve a consistent solution.

Because of the paradox derived from interpreting the composition of mainstream SiC grains as the consequences of variations in the metallicity of the parent stars due to temporal GCE, more recent interpretations have proposed solutions that avoid a strict temporal relationship between the Sun and the grain parent stars. Clayton [70] proposed an interpretation of the Si composition of SiC grains based on the fact that AGB stars during their life could have changed their position with respect to their distance from the Galactic centre. Stellar orbital diffusion is caused by the granular and time-dependent gravitational potential well of the Galactic disk, in particular because of the presence of massive molecular clouds and/or spiral arms, but it is difficult to model. Wielen, Fuchs, & Dëttbarn [293] advocated that the Sun itself formed closer to the Galactic centre and has diffused outward to the present location. The AGB stars, parents of SiC grains, could also have diffused from birthplaces closer to the Galactic centre to the location where the Sun was born, where they shed the SiC grains. Because of the presence of a metallicity gradient within the Galactic radius, the metallicity of stars born closer to the Galactic centre is higher, and hence the silicon isotopic ratios of these stars are higher than solar. This interpretation is fascinating in its relating of the composition of mainstream SiC grains to dynamical properties of the Galaxy. However, the quantitative semianalytic analysis of Nittler & Alexander [208] showed that the orbital diffusion model would not predict that most presolar grain parent stars had metallicities higher than solar. More recently, it has been proposed that the same effect of orbital diffusion is produced by scattering from spiral waves [251], i.e. density waves of unknown origin moving through the Galactic disk that are believed to produce the Galactic spiral structure.

Another effect that could contribute to the observed distribution of the Si composition in SiC grains was proposed by Lugaro *et al.* [177], and involves inhomogeneities in the interstellar medium from which the grain parent stars were born. Inhomogeneities can be produced by the stochastic nature of the mixing of ejecta from supernovæ of different types and masses, thus creating regions of the interstellar medium of slightly different composition at the same time and at the same distance from the Galactic centre. This model produced interesting results, but needs more investigation, in particular in relation to the observed variations in the Ti isotopic composition (see next section). In any case, this approach creates the possibility of constraining the degree of inhomogeneity in the Galaxy to a higher level than is possible by spectroscopic observations [207].

Another interpretation of the mainstream line has been proposed by Clayton [65]. In this model the features of the Si composition in SiC grains are attributed to the fact that the AGB parent stars of SiC grains could have all formed in a starburst generated by the merger, about 6 Gyr ago, of a metal-poor satellite galaxy with our Galaxy. This model is also preliminary and needs more testing.

In conclusion it is premature to decide which, if any, of the above explanations is the correct answer for the Si isotopic composition of mainstream SiC grains, and the attempts to decode the information hiding in the mainstream line have to be considered far from complete.

4.7 Titanium isotopic composition of mainstream SiC grains

Titanium data for $\simeq$ 90 single mainstream SiC, A+B and Y grains are shown in Fig. 4.10 [3, 14, 15, 122]. Isotopic ratios are represented as permil variations with respect to solar, $\delta(^i\mathrm{Ti}/^{48}\mathrm{Ti})$. The isotope taken as reference is the abundant $^{48}\mathrm{Ti}$, which represents 74% of the total titanium in the solar system. Titanium-46 and $^{47}\mathrm{Ti}$ represent about 8% of solar titanium, respectively, while $^{49}\mathrm{Ti}$ and $^{50}\mathrm{Ti}$ share the remaining 10% in equal parts. Error bars for $\delta(^{29}\mathrm{Si}/^{28}\mathrm{Si})$ are within the size of the symbols in the plot, while, unfortunately, error bars for the titanium data are typically quite large, especially for the A+B grains. As shown in Fig. 4.10 SiC grain data show large excesses only in $^{50}\mathrm{Ti}$, with $\delta(^{50}\mathrm{Ti}/^{48}\mathrm{Ti})$ up to $\simeq 400^o/_{oo}$. Moderate excesses are present in $\delta(^{46}\mathrm{Ti}/^{48}\mathrm{Ti})$ and $\delta(^{49}\mathrm{Ti}/^{48}\mathrm{Ti})$, up to $\simeq 200^o/_{oo}$, and small excesses are shown by $\delta(^{47}\mathrm{Ti}/^{48}\mathrm{Ti})$, up to typically $\simeq 100^o/_{oo}$. This is qualitatively consistent with the fact that neutron-capture nucleosynthesis in AGB stars mostly affect the $^{50}\mathrm{Ti}/^{48}\mathrm{Ti}$ ratio, because $^{50}\mathrm{Ti}$ has a magic number of neutrons, thus a small neutron-capture cross section and tends to accumulate during the neutron flux, and to a lesser

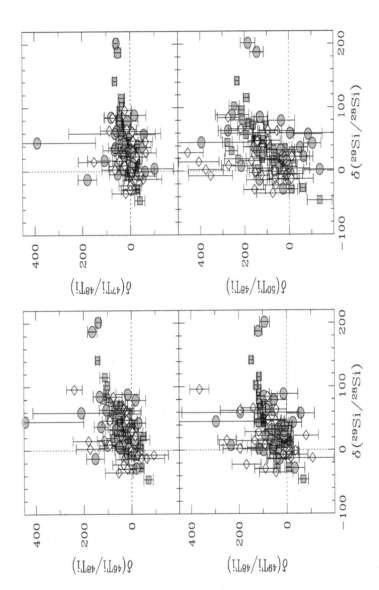

Fig. 4.10 Titanium isotopic ratios measured in mainstream (squares), A+B (circles) and Y (diamonds, symbols as in Figs. 4.1 and 4.2) single SiC grains as a function of the $^{29}Si/^{28}Si$ isotopic ratio. Since the usual δ notation is employed, as in Fig. 4.9, the solar composition corresponds to the origin of the plot and it is shown by the dotted lines. One SiC-Y grain has $\delta(^{50}Ti/^{48}Ti) = 990^{\circ}/_{\circ\circ}$ and lies outside the range of the plot.

extent the ^{49}Ti/^{48}Ti and ^{46}Ti/^{48}Ti ratio, while the ^{47}Ti/^{48}Ti ratio is left almost unchanged [177].

The titanium isotopic ratios appear to correlate with the silicon isotopic ratios, hence these data should also be taken into account when looking for an explanation of the puzzle of the Si composition of mainstream SiC grains described in the previous section. However, the GCE trend of titanium isotopes presents many problems [279]: not enough ^{48}Ti is produced by supernovæ, ^{47}Ti is largely underproduced, and the neutron rich isotope ^{50}Ti seems to need a peculiar site of production, maybe a special type of SNIa [295]. Considering all the difficulties related to the Galactic origin of Ti and the fact that only a small set of SiC grain data is available, it is not surprising that the composition of titanium in SiC grains still represents a problem even more puzzling than that of Si. Only after enlarging the dataset and improving stellar models can we hope to better understand the information carried by titanium in presolar SiC grains.

Thanks to the NanoSIMS, titanium data are now available also for $\simeq$ 10 SiC-Z grains. SiC-Z grains are more abundant among smaller grains and hence are more easily recovered using the new instrument. The $\delta(^{46,47,49}$Ti/^{48}Ti) of SiC-Z grains are mostly negative, down to $-300^o/_{oo}$ and correlate with $\delta(^{29}$Si/^{28}Si). This is in agreement with the interpretation that such grains have been produced in AGB stars of low metallicity (see Sec. 4.8.1). In fact, similarly to the case of the Galactic evolutionary trend of the Si isotopic ratios discussed in Sec. 4.6, the ^{46,47,49,50}Ti/^{48}Ti ratios decrease with metallicity [279] because ^{48}Ti is a *primary* nucleus, while ^{46,47,49}Ti are of *secondary* production. Only the $\delta(^{50}$Ti/^{48}Ti) observed in the grains are typically positive or close to zero, which is explained by the effect of AGB nucleosynthesis.

Other intermediate-mass elements, such as Ca, Cr and Fe have been measured in SiC grains in bulk, hence yielding average information [18]. The possibility of applying NanoSIMS and RIMS (see Chapter 3) to the measurements of the composition of intermediate-mass elements in single SiC grains is another future exciting prospective, both for further testing nucleosynthesis processes in AGB stars, and for the opportunities that such data would represent in yielding the most detailed information on the Galactic chemical evolution of each of the isotopes of intermediate-mass elements.

4.8   A, B, X, Y and Z: The minor SiC grains populations

4.8.1   *The Y and Z populations*

Like mainstream SiC grains, these two small populations are believed to have formed in the extended envelopes of C-rich AGB stars. However, unlike mainstream grains, the metallicity of the parent stars of these grains is believed to be lower than solar: about half solar in the case of Y grains [14] and about one third of solar in the case of Z grains [124]. In AGB stellar models, when the metallicity is decreased, the temperature at the base of the thermal instabilities increases. In this way, more neutrons are released by the ^{22}Ne$(\alpha,\mathrm{n})^{25}$Mg neutron source and the composition of silicon is more strongly affected than in AGB models of solar metallicity. In particular, the ^{30}Si/^{28}Si ratio significantly increases, with respect to solar, thus matching the enhancements observed in SiC-Y and SiC-Z grains. The ^{29}Si/^{28}Si ratio, on the other hand, is not much altered by nucleosynthesis in AGB stars and can be used as a tracer of the initial composition of the star. By combining detailed theoretical nucleosynthesis models and high-precision data for SiC grains of the mainstream, Y and Z populations, it is possible in principle to reconstruct the Galactic chemical evolution of the Si isotopes down to metallicities of approximately one third of the solar value [303].

The C and N composition of the grains must also be placed in such a framework. SiC-Y grains have ^{12}C/^{13}C ratios higher than mainstream grains: they are, in fact, classified as having ^{12}C/^{13}C > 100. This feature is in agreement with predictions of AGB stellar models with low metallicity, where the efficiency of mixing by third dredge-up increases so that more ^{12}C is carried to the surface and the ^{12}C/^{13}C ratio is increased. However, grains belonging to the Z population behave in the opposite way: their ^{12}C/^{13}C ratios are typically similar or lower than those of the mainstream grains. This feature has been interpreted as an indication that extra-mixing processes, converting ^{12}C into ^{13}C at the base of the convective envelope (see Sec. 4.3), must have been more efficiently at work in the parent stars of SiC-Z grains [124, 210].

As more grains of type Y and Z are discovered thanks to improved techniques, and more data are progressively available on other elements present in the grains (see e.g. Sec. 4.7), it will be possible to better explain these populations in our theoretical framework. This will allow us to extract precious information on AGB stars of a range of metallicities and on the Galactic chemical evolution of the isotopes of the elements present in the grains.

4.8.2 The A and B populations

The origin of SiC grains of type A and B is still a mystery. The low $^{12}C/^{13}C$ ratios are a signature of H burning, which is also shown by the fact that A+B grains have, on average, higher $^{26}Al/^{27}Al$ ratios than mainstream grains. Amari *et al.* [15] proposed that the most likely sources of at least a fraction of A+B grains are C(J) stars: carbon stars of type J, which show $^{12}C/^{13}C$ ratios around 3, the equilibrium value of H burning. However, the nature of these stars themselves is not understood! Stars evolving after the AGB phase on the white-dwarf cooling track, but undergoing a very late thermal pulse, *born-again* AGB stars, have also been invoked as a possible source. In any case, the A+B grains with $^{14}N/^{15}N$ ratios much lower than solar remains unexplained. One suggestion is that the rate of the $^{18}O(p,\alpha)^{15}N$ reaction should be 1,000 times higher that the current estimate [132]. However, uncertainties for this rate appear to be within a factor of four [23]. Another scenario is that the CN composition of at least a fraction of A+B grains is related to nova nucleosynthesis occurring on CO white dwarves [176], thus possibly relating these grains to the smaller family of the nova grains, whose origin has been attributed to nova nucleosynthesis occurring on more massive NeO white dwarves [7].

While for the Si and Ti compositions A+B grains cover approximately the same range as mainstream grains (even though their Si correlation line appears to have a slope of $\simeq 1.5$, instead of $\simeq 1.3$ for the mainstream grains), they show different features with regard to the presence of an *s*-process signature in the composition of heavy elements. The concentrations of heavy trace elements in A+B grains show a range of patterns [10]. Out of 21 A+B grains that were analysed for their trace element abundances, two thirds of them showed a heavy element distribution compatible with an enrichment of *s*-process elements. However, among the twelve single A+B grains analysed so far with RIMS for their Mo, Ba and Zr isotopic compositions, not one grain was found with the *s*-process signature. One grain was found to show the signature of a SNII neutron burst (similarly to SiC-X grains, see next section) and one grain showed the signature of the *p* process [244]! These recent data make the mystery of A+B grains even more intriguing.

4.8.3   *The X population*

SiC-X grains are enriched in ^{15}N, depleted in [29]Si and [30]Si and have high
[26]Al/[27]Al ratios. Many of them also show excesses of [12]C. About 10 – 20%
of the grains have [44]Ca excesses, and some grains also have [49]Ti excesses.
The Ca and Ti anomalies were most likely produced by the decay of radioactive nuclei [49]V (half-life = 330 days) and [44]Ti (half-life = 50 years) and
clearly point toward supernovæ as the site of origin of these grains, since
these isotopes are only synthesised in supernovæ [8, 31, 125, 127, 129, 214].
The [44]Ti excesses are correlated with those of [28]Si, which indicates that the
grains contain material from the inner [28]Si-rich regions of SNII (Fig. 2.5)
where [44]Ti can be produced by alpha-rich freeze out (Sec. 2.4). Silicon
nitride grains have typically the same isotopic signatures as SiC-X grains
and, although [44]Ca excesses have so far not been detected, their origin is
also attributed to SNII [215].

Even though SNIa have also been proposed as the source of SiC-X grains
[66], it appears now more likely that most of these grains formed in SNII.
Extensive mixing of the different layers ejected by a SNII is needed to match
the composition of presolar grains from supernovæ. In fact, [28]Si and [44]Ti
are produced in the inner regions of the exploding star, while low ^{14}N/[15]N
and high [26]Al/[27]Al ratios are produced in the He-burning and H-burning
layers, respectively. Different types of mixtures have been calculated and
compared to supernova grain data by Travaglio *et al.* [281] for graphite
grains (see Sec. 6.2), Hoppe *et al.* [129] for SiC-X grains and Yoshida &
Hashimoto [299] for both graphite and SiC-X grains. Problematic isotopes
seem to be ^{15}N and [29]Si, which appear to be typically underproduced in
SNII with respect to the grain data. Actually, there is also an indication
for two different populations of SiC-X grains: one with higher-than-solar
and one with lower-than-solar [29]Si/[30]Si ratio [170]. The problem of the Si
composition of SiC-X has been addressed also by Yoshida *et al.* [300].

One problem is that, to satisfy the C/O > 1 condition for the formation of carbonaceous dust, the contribution of material from the middle
oxygen-rich layers of the star (see Fig. 2.5) should be very limited, yet allowing material from the inner and from the outer regions to get well mixed.
Clayton and coworkers [67, 69] have shown that carbon-rich dust, such as
graphite and SiC, could condense in supernova environments even out of
O-rich gas, even though Ebel & Grossman [88] have disputed such a claim,
at least for SiC. In any case, because of this possibility, the constraints on
the C/O ratio of the mixture required to match the composition of SNII

grains could be relaxed. A study of the formation of SiC in a SNII environment together with the effects of implantation of material into the grains while they pass through different regions has been carried out by Deneault *et al.* [81].

Thanks to RIMS (Sec. 3.3.3), Pellin and collaborators [223] have analysed the composition of Mo and Zr in several SiC grains of type X producing further detailed insights into the nucleosynthetic processes in SNII that involve the production of heavy elements. These compositions cannot be explained by the occurrence of pure *s*-process or pure *r*-process nucleosynthesis. A possible scenario for reproducing the observed compositions involves a short and intense neutron burst [193], which, interestingly, is also one of the scenarios invoked to explain the composition of Xe-H associated to presolar diamonds, described in Sec. 6.1. This interpretation provides a link between the different presolar grains of supernova origin and could also shed some light on the understanding of neutron fluxes in supernovæ and the difficult task of finding the *r*-process source.

4.9   Exercises

(1)   a) Which fraction of He intershell material mixed into the envelope by third dredge-up (TDU), in other words which dilution $DIL = M_{TDU}/M_{env}$, is necessary to achieve C/O > 1 in a star of solar metallicity? Consider that the mass fraction of a given isotope i in the envelope during the TDU phase can be calculated as:

$$X^i = \frac{X^i_{initial} M_{env} + X^i_{intershell} M_{TDU}}{M_{env} + M_{TDU}},$$

and use as initial ^{12}C the abundance modified by the first dredge-up (FDU): $X^{FDU}_{12} = 0.002$ and for ^{16}O the solar abundance: $X^{\odot}_{16} = 0.009$. The typical He-intershell values are $X^{intershell}_{12} = 0.23$ and $X^{intershell}_{16} = 0$.

b) Which dilution is needed if the stellar metallicity is one third of the solar value? (In this case simply divide by three the initial values given above). Comment on the result.

c) Show that for the dilutions calculated above the $^{12}C/^{13}C$ ratio is equal to 36 for both solar and one third of solar metallicity. (Start with a ^{13}C abundance as modified by the first dredge-up and possible extra-mixing process: $X^{FDU+extra\,mixing}_{13} = 0.0002$

and consider that all ^{13}C in the He intershell is destroyed by α captures.)

(2) Calculate the values of the ^{4}He/^{22}Ne and ^{20}Ne/^{22}Ne ratios in the He intershell of AGB stars of different metallicities using the simple method described in Sec. 4.4. (Note that $X_{20}^{\odot} = 0.00162$)

(3) a) Timmes & Clayton [278] showed that because of GCE effects the silicon isotopic ratio with respect to solar: $\dfrac{^{29}\mathrm{Si}/^{28}\mathrm{Si}}{(^{29}\mathrm{Si}/^{28}\mathrm{Si})_{\odot}}$ varies roughly as the square root of the metallicity of the star with respect to solar: $\sqrt{Z/Z_{\odot}}$. What is the range of metallicities for the mainstream SiC grains parent stars if we assume that the variations in the Si isotopic composition are due to GCE effects?

  b) Calculate a simple average of the metallicity of mainstream, Y and Z SiC grains. How do they stand with respect to each other? Compare the result with the discussion on Y and Z grains of Sec. 4.8.1.

  c) We can use the following approximate relationship between the metallicity of the star with respect to solar and the time when the star was born: $Z/Z_{\odot} = 0.086(t + 1)$, where t is the time from the birth of the Galaxy expressed in billions of years. If the Sun was born at $t = 10$ billion years, when were the parent stars of mainstream SiC grains born with respect to the Sun? Are these results realistic?

Chapter 5

Heavy Elements in
Presolar SiC Grains

Elements heavier than Fe are present in *trace* amounts in presolar SiC grains. Their abundances are typically of the order of a few parts per ten thousand in mass (i.e. 100 ppm) [10], however, large variations are observed. Different abundances are observed depending on the grain size, since smaller grains have higher abundances, and on the specific element considered, since more refractory elements, such as Zr, are typically present with higher abundance than less refractory elements, such as Sr. Even individual grains exhibit a variety of abundances. In fact, nine different patterns were introduced by Amari *et al.* [10] to classify the variations in the abundance distribution of heavy elements observed in 60 individual SiC grains of large size, belonging to the KJH sample (Fig. 3.1). This multiplicity of abundance patterns is observed in grains all belonging to the mainstream SiC population. The different patterns are interpreted using thermodynamic equilibrium calculations of the condensation of dust in stellar atmospheres, and considering how heavy elements are trapped in the dust [173]. Moving from more refractory to more volatile elements, the mechanism of inclusion in SiC grains shifts from condensation, which results in higher abundances, to ion implantation, which results in lower abundances. The heavy noble gases, Kr and Xe, which are the most volatile, and thus the least abundant heavy elements in SiC grains, will be discussed separately in the last section of this chapter.

The clearest feature of the abundance distributions of heavy trace elements in mainstream SiC grains is the signature of the *s* process. Elements produced by the *s* process, such as Zr and Ba, are typically enhanced in SiC grains, which implies that in the gas from which the grains formed *s*-process elements were up to a factor of 30 overabundant with respect to solar. This is in agreement with the fact that SiC grains are the carriers of the isotopically anomalous Xe-S component (see Sec. 1.2 and Fig. 1.3), which shows the signature of the s process. All the isotopic ratios of heavy elements in mainstream SiC grains show a clear s-process signature, with enhancements in the s-only isotopes, for example ^{128}Xe and ^{130}Xe in the case of Xe, and ^{134}Ba and ^{136}Ba in the case of Ba. As discussed in the previous chapter, this signature is an important piece of evidence relating the origin of mainstream SiC grains to carbon-rich Asymptotic Giant Branch stars, which are observed to have s-process enriched envelopes.

Given such strong evidence that the composition of the material present in mainstream SiC grains was modified by the s process, and since the parameters that affect the s-process abundances (e.g. the temperature and the neutron density) are model dependent, grain data can be used to better constrain both the origin of the grains (for example the range of masses and metallicities of the parent stars) and the modelling of the s process in AGB stars. Data on the composition of heavy elements in presolar grains are given in the form of isotopic ratios with error bars typically of 5% to 20% for single grain analysis, and much lower for analysis of grains in bulk. This kind of detailed and precise information is not achievable from the spectroscopic observations of stellar atmospheres and makes presolar SiC grains the best tool for testing s-process stellar models. Single SiC grain data allow us to recognise even very subtle differences among the way the s process occurs in different stars, since many different stars produced the SiC grains studied in the laboratory. Hereafter, the current model for the s process in AGB stars is described (see also [49] for a review), together with the information that can be extracted from the isotopic composition of heavy elements in SiC grains.

5.1 Modelling the s process in AGB stars

In the He-rich intershell of AGB stars the temperature and the number of α particles are high enough to trigger α-capture reactions, among which are the (α, n) reactions that result in the production of neutrons. The structure of an AGB star is shown in Fig. 2.4 and its evolution in time has been described in Sec. 4.2.1. The maximum temperature in the star, of the order of a few hundred million degrees, is reached in the He intershell. The region is rich in α particles because in the H-burning shell all H is transformed into ^{4}He, which in turn is only partially destroyed by the recurrent activation of the He-burning shell. Of the He intershell typically 70% by mass is made up of α particles.

Spectroscopic observations indicate that efficient s processing must occur in AGB stars, however, exactly how, where and when the neutrons are produced is still a matter of debate. As described below, this search involves the study of complex features of stellar evolution such as the treatment of mixing, in particular at convective boundaries, the role of stellar rotation and the effect of magnetic fields on the structure and evolution of stars. The research is guided by observational constraints, among which are those produced by SiC grain measurements.

5.1.1   *The neutron source in AGB stars*

Two nuclei represent the best candidates for the source of neutrons in the intershell of AGB stars: ^{13}C, which produces neutrons via the ^{13}C$(\alpha, n)^{16}$O reaction, and ^{22}Ne, which produces neutrons via the ^{22}Ne$(\alpha, n)^{25}$Mg reaction. The ^{22}Ne$(\alpha, n)^{25}$Mg reaction is efficiently activated at temperatures higher than approximately 3×10^8 K. The rate of this reaction (see Exercise 2.2 for a definition) is plotted in Fig. 5.1, together with the rate of the ^{13}C$(\alpha, n)^{16}$O reaction. The ^{13}C$(\alpha, n)^{16}$O reaction is activated at much lower temperatures, around 0.9×10^8 K, than the ^{22}Ne$(\alpha, n)^{25}$Mg reaction.

In early studies, ^{22}Ne was considered to be the main neutron source for the s process [137, 138, 283]. As discussed in Sec. 4.4, a large abundance of ^{22}Ne is present in the He intershell, since all the initial CNO abundances are converted into ^{22}Ne. Stellar models show that the maximum temperature in AGB stars is reached at the base of the convective regions produced by the thermal pulses in the He intershell. Models of low-mass AGB stars (lower than about 5 $M_\odot$) barely reach temperatures as high as 3×10^8 K, and only for very short times (a few years) during the thermal pulses. Hence, the s process driven by the activation of the ^{22}Ne neutron source can only be efficient in intermediate-mass AGB stars (higher than $\simeq 5$ $M_\odot$), where the maximum temperature reaches $\simeq 3.8 \times 10^8$ K.

However, in the early 1990s, observational evidence from data on AGB stellar luminosities [94] and kinematics [90], came by indicating that the chemically peculiar giant stars showing s-process enhancements are AGB stars of low mass. Moreover, the efficient activation of the ^{22}Ne neutron source in AGB stars would predict abundance features that were not observed in these stars. For example, the ^{22}Ne$(\alpha, n)^{25}$Mg and ^{22}Ne$(\alpha, \gamma)^{26}$Mg reactions, which are activated at similar temperatures, would produce large excesses in the heavy Mg isotopes, ^{25}Mg and ^{26}Mg. However, the abun-

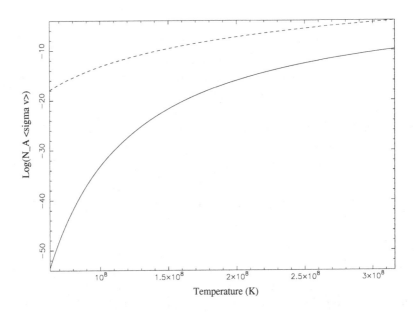

Fig. 5.1   Reaction rates $N_A \langle \sigma v \rangle$ (cm^3 s^{-1} mole $^{-1}$), where N_A is Avogadro's number and $\langle \sigma v \rangle$ is the Maxwellian average of the cross section of the reaction times the
relative velocity, as a function of the temperature for the neutron-producing reactions
$^{13}C(\alpha, n)^{16}O$ (dashed line) and $^{22}Ne(\alpha, n)^{25}Mg$ (solid line) from [23]. As a rule of thumb,
a reaction starts to operate at a temperature where its rate is above $\simeq 10^{-15}$ cm^3 s^{-1}
mole^{-1}. However, to properly determine the actual occurrence of a given reaction the
conditions at which it operates have to be taken into account. The actual number of
reactions will vary in different conditions depending on the availability of the interacting
nuclear particles, the density of the material, and the time-scale defined by how long the
temperature is high enough to activate the reaction (see Exercise 5.1).

dances of these isotopes in s-process enriched AGB stars are similar to
those in the Sun [182, 254]. Moreover, s-process models with the ^{22}Ne
neutron source activated at high temperatures during thermal pulses in
intermediate-mass AGB stars, produce high neutron densities. This feature
results in abundances, for some isotopes, incompatible with the abundances
in the solar system, particularly for isotopes produced by the activation of
branching points on the s-process path (Sec. 2.5.1) such as ^{96}Zr [85]. With
regards to the formation of SiC grains, it is unlikely that the bulk of the
grains had formed in intermediate-mass AGB stars since the formation of
carbon-rich stars in this mass range is inhibited by the occurrence of hot
bottom burning (Sec. 4.2.1).

Theoreticians thus turned to the other possible neutron source: the ^{13}C

nuclei. In this case the situation is reversed with respect to the ^{22}Ne neutron source: the temperature in low-mass AGB stars is high enough to efficiently activate the ^{13}C$(\alpha, n)^{16}$O reaction, however, there is not enough ^{13}C in the intershell to release a neutron flux capable to produce the observed s-element enhancements. In fact, the mass fraction of ^{13}C in the H-burning ashes produced by the CNO cycle is very small, $X_{13} \simeq 3 \times 10^{-5}$. This results in a total mass of ^{13}C in the He intershell, at the onset of thermal instability, of $M_{13} \simeq 1.5 \times 10^{-7} \ M_{\odot}$, since the H-burning ashes have a mass $M_{ashes} \simeq 5 \times 10^{-3} \ M_{\odot}$. An amount of ^{13}C in mass, on average, of the order of a few $10^{-6} \ M_{\odot}$ is needed in the He intershell to produce neutrons fluxes that would result in heavy-element enhancements matching the observed composition of AGB stars. This is one order of magnitude higher than the amount available from the H-burning ashes.

5.1.2 *The production of a* ^{13}C *pocket*

In the current model [97, 104, 266], which is schematically represented in Fig. 5.2, it is assumed that some partial mixing between the H-rich envelope and the He intershell occurs, in order to produce a region containing the extra amount of ^{13}C needed for the s process. This region is generally referred to as the ^{13}C *pocket*. The abundant ^{12}C present in the He intershell ($\simeq 23\%$ in mass fraction) captures the protons that are assumed to have been mixed down from the envelope, producing ^{13}N, which quickly decays into ^{13}C. If the mass fraction of protons is higher than about 1%, then ^{13}C is also destroyed by proton capture, resulting in the production of ^{14}N, as in the CN cycle. With an initial abundance in number per atomic mass unit of protons of $Y_p = 0.007$, which is a favourable situation for the production of ^{13}C, we can assume that all the protons are captured by ^{12}C to produce ^{13}C. Hence, the abundance by number of the ^{13}C produced is $Y_{13} = 0.007$, which corresponds to a mass fraction $X_{13} \simeq 0.09$. Thus, the mass of the pocket must be $M_{pocket} \simeq 10^{-5} \ M_{\odot}$ in order to obtain the needed amount in mass of ^{13}C: $M_{13} = X_{13} \times M_{pocket} \simeq 10^{-6} \ M_{\odot}$.

Mixing of material from the proton-rich envelope with the ^{12}C-rich intershell is not possible during the interpulse period because the H-burning ashes are located between these two regions. During a thermal pulse, because of the developing of the convective region in the whole intershell, ^{12}C-rich material is pushed outwards reaching close to the envelope. However, it was shown early on that mixing does not occur between the convective pulse and the convective envelope because the H-burning shell is still active

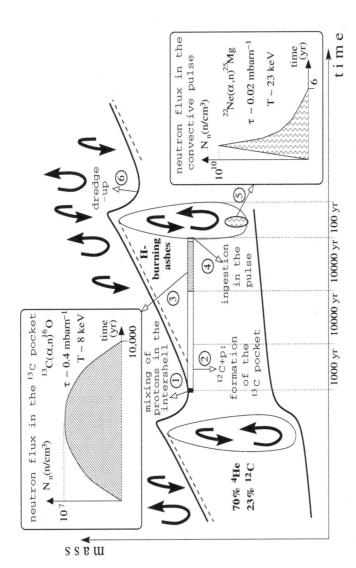

Fig. 5.2   Schematic representation of the current model for the *s* process in the He intershell of AGB stars (a variation of Fig. 4.4). The six separate mixing and nucleosynthesis phases leading to enrichment of *s*-process elements at the stellar surface (1: proton diffusion, 2: ^{13}C production, 3: activation of the ^{13}C neutron source, 4: mixing in the convective pulse, 5: activation of the ^{22}Ne neutron source, and 6: third dredge-up) are indicated. Rough time-scales of the nucleosynthetic phases in the ^{13}C pocket and of the convective pulse are reported on the x-axis.

at that time, making up a barrier to mixing [139]. This was confirmed by all the recent models of AGB stars[1]. A more favourable time for the mixing of protons into the intershell can be identified as the moment when, after the end of each third dredge-up episode, the H-burning shell is temporarily extinguished and the H-rich convective envelope and the He-rich radiative intershell make contact. Quickly after this time the convective envelope retreats, and a "buffer" radiative H-rich region, which will remain during the whole interpulse period, is formed between the envelope and the He intershell preventing further mixing.

The exact mechanism by which the penetration of protons in the intershell occurs is still a matter of debate. Stellar models usually identify convective regions using the Schwarzschild criterion. This is essentially based on Archimedes' law that a body in a fluid is subjected to a buoyant force equal to the gravitational force of the fluid displaced. Let us assume that a "blob" of gas in a star makes a small vertical movement under adiabatic conditions. (We can assume this occurs adiabatically since heat exchange typically takes more time than the movement, but it can be shown that the conclusions below are valid also in non-adiabatic conditions). Let us imagine, for example, that the motion occurs towards a hotter region of the star. If the temperature variation of the blob in space $|dT/dr|_{adiabatic}$ is smaller than the temperature variation of its surroundings $|dT/dr|_{radiative}$, then the element becomes cooler and more dense than its surroundings and will continue to sink. The Schwarzschild criterion for stability against convection is hence expressed by $|dT/dr|_{adiabatic} > |dT/dr|_{radiative}$, or more usually in the form, $\nabla_{adiabatic} > \nabla_{radiative}$, where $\nabla = (d \log T)/(d \log P)$ and P is the pressure, noting that dP/dr is the same for the element and its surroundings since pressure imbalances are efficiently removed by sound waves.

Since $|dT/dr|_{radiative}$ is a function of the opacity of the material, and thus of its composition, special situations can arise when the change in the composition produced by mixing makes the region radiatively stable, and thus stops the mixing. This is called semiconvection. In particular, the effect on opacity of recombination of carbon nuclei and electrons at the top of the intershell region, was shown to drive semiconvection below the base of the convective envelope during third dredge-up, and for the first time to produce self-consistent mixing of protons and ^{12}C, thus leading to the formation of a ^{13}C pocket in stars of low metallicity [141].

---

[1]Except for models of AGB stars of very low metallicy [144].

Within the Schwarzschild criterion, by definition, an element in a convective region is always affected by forces that favour its motion, and the shift from motion to stability at the border between convective and radiative regions must occur in the radiative regions (*convective overshoot*). Hence, within the framework of the Schwarzschild criterion it is physically reasonable to imagine that in AGB stars some protons from the base of the convective envelope could penetrate into the radiative intershell region. Whether this effect is large enough to produce the needed ^{13}C is still uncertain. It must also be noted that during third dredge-up, at the base of the convective envelope, the application of the Schwarzschild criterion can be rather ambiguous. This is due to the fact that the different compositions, of the envelope and of the intershell, produce a discontinuity in the opacity of the material, which is used to calculate the quantities that define the Schwarzschild criterion. Thus a discontinuity results in the $\nabla_{adiabatic} - \nabla_{radiative}$ function, which makes it difficult to define the convective border, and leads to large uncertainties not only in the treatment of a possible proton diffusion, but also in the calculation of the amount of third dredge-up. This is one of the main uncertainties present in AGB models [95, 175, 197].

Many studies now deal with the limitations implicit in the use of the Schwarzschild criterion. Multi-dimensional simulations of shallow convective regions have shown that the decrease of the diffusion coefficient, which measures the rate of mixing, or, equivalently, of the velocity of the material in radiative regions in contact with a convective border, can be represented by an exponential law [93]. These types of simulations are computationally expensive and hence at the moment cannot by applied to the large convective envelopes of AGB stars. However, the convective borders can be modified within conventional one-dimensional stellar models in order to model diffusion of material from convective into radiative regions with exponentially decreasing velocities. This type of treatment of the convective borders does lead to the mixing of protons in the He intershell and hence the formation of a ^{13}C pocket [115, 117]. However, the extent in mass of the mixing region and hence the total amount of ^{13}C produced, is a function of a free parameter, f, known as the *overshoot* parameter, which governs the rate of the exponential decay of the velocity in the radiative zone adjacent to the convective border. The larger the overshoot parameter, the deeper the mixing of protons, and hence the larger the mass of the ^{13}C pocket. Only if f is set to a value $\simeq 0.1$ is the mass extent of the mixing zone large enough to result in the amount of ^{13}C needed.

Very similar results are obtained when the effect of *internal gravity waves* is considered at the border between the convective envelope and the radiative region. These are small oscillations around equilibrium of material in radiative regions. It is hypothesised that the convective cells beating against the bottom of the convective zone drive gravity waves. The waves would then break at the boundary like ocean waves and allow mixing to occur across the border. When this type of mixing is taken into account a ^{13}C pocket of mass of the order of a few 10^{-4} $M_\odot$ is produced [82].

Another possibility is the inclusion of rotation in stellar models. All stars rotate, and rotation produces changes to the stellar structure, in particular to the location of mixing zones with respect to non-rotating models. When rotation is included in AGB stellar models, shear mixing occurs between the faster-rotating core and the slower-rotating convective envelope. This shear mixing can account for the production of a region in which protons and intershell material are mixed, leading to the formation of a ^{13}C pocket of very low mass, of the order of 10^{-6} $M_\odot$ [161]. However, it was also recognised that rotationally-induced mixing actually continues during the interpulse period, hence inhibiting the production of heavy elements by the s process [118, 253]. This effect and its consequences will be discussed in more detail in Sec. 5.2.3.

5.1.3 *The current model*

Once assumed that the formation of the ^{13}C pocket occurs, the features of the s process depend on the thermodynamical and structural properties of AGB stars. The temperature in the ^{13}C pocket increases during the interpulse period, typically reaching 100 million degrees before the onset of the next thermal pulse. By this time, all the ^{13}C is consumed via (α, n) reactions, neutrons are released, and the ^{13}C pocket is turned into an s-process pocket, where s-process elements are overproduced by up to three orders of magnitudes compared to their initial abundances. The layer is then engulfed in the convective zone driven by the thermal instability and mixed with the rest of the He intershell material (see Fig. 5.2). The s-process material is thus diluted by a factor of roughly 1 part per 10 with material from H-burning ashes (with solar abundances of elements heavier than Fe) and roughly another 1 part per 10 with material from the He intershell, carrying the signature of the previous s-process nucleosynthesis. This mixing and dilution mechanism ensures that the s-process material is spread all throughout the intershell, and thus carried to the surface by the following third dredge-up episode.

During the thermal pulse a second neutron flux occurs, produced by the ^{22}Ne$(\alpha, n)^{25}$Mg reaction, which is marginally activated at the bottom of the convective zone when the temperature is above 2.5×10^8 K. After the quenching of the thermal pulse, the subsequent third dredge-up episode is responsible for mixing the s-processed material to the surface, where it is observed and where it can be included in SiC grains.

It is instructive to compare the features of the two neutron fluxes: that produced by the ^{13}C source, and that produced by the ^{22}Ne source, which are described in detail in the next two subsections. The neutron densities as a function of time generated by each neutron source are schematically presented in Fig. 5.2. They appear to have very different features from each other and hence affect the final s-process abundance distribution in different ways. The neutron flux in the ^{13}C pocket is produced during the period in between pulses, at a typical temperature of 0.9×10^8 K ($\simeq 8$ keV), and under radiative conditions, which means that there is no interaction between the different layers of the ^{13}C pocket. This has been assumed so far when computing the s process in the ^{13}C pocket, however, it is not the case if stellar rotation is included in the computation of the stellar structure, as will be discussed in detail in Sec. 5.2.3. The burning of ^{13}C and hence the neutron release occurs slowly, on relatively long time-scales of the order of 10^4 years. Hence, the neutron density at any given time is relatively low, reaching a maximum of about 10^7 cm^{-3} in low-mass AGB stars of solar metallicity. Under such conditions, no branchings are open on the s-process path (Sec. 2.5.1). The time-integrated neutron flux, or neutron exposure as defined in Sec. 2.5.1, is quite large, on the order of 0.4 mbarn^{-1}, which results in the production of large overabundances of the s-process elements.

On the other hand, the neutron flux produced by the ^{22}Ne source is released within the convective pulse, at a typical temperature of 2.7×10^8 K ($\simeq 23$ keV), under convective conditions and can be well described as intense and brief. The evolution of the neutron density follows that of the rate of the ^{22}Ne$(\alpha, n)^{25}$Mg reaction, which in turn follows the evolution of the temperature at the base of the convective pulse. A thermal pulse is a quasi-explosive instability, which lasts for a short time, but reaches high temperatures. The neutron flux lasts for only a few years and thus the neutron exposure is approximately 20 times smaller than that produced in the ^{13}C pocket. However, the neutron density reaches values as high as 10^{10} cm^{-3}, in low-mass AGB stars of solar metallicity, which can activate branchings on the s-process path.

In summary, the ^{13}C neutron source is responsible for the production of the bulk of *s*-process elements in low-mass AGB stars, while the activation of the ^{22}Ne neutron source contributes to the final abundance distribution, chiefly by modifying the abundances of isotopes affected by branchings.

5.1.4   *The neutron flux in the ^{13}C pocket*

Even though the mechanism by which the ^{13}C pocket is formed is still a matter of debate, it is possible to constrain some properties of the nucleosynthesis occurring in the pocket by considering the activation of nuclear reactions following the assumption that some protons have entered the He intershell. It is reasonable to imagine that the protons entering the He intershell are characterised by a distribution in which the abundance of protons decreases somehow continuously with the depth in mass, from the maximum envelope value, where protons represent about 70% of the mass, to zero. In the following we also assume that, after the protons have been mixed into the He intershell, no further mixing occurs in the ^{13}C pocket, and thus nucleosynthesis processes in the different layers of the pocket occur independently of each other and are only governed by the initial amount of protons in each layer. Since the question of the specific shape of the proton profile as a function of the depth in mass is still open, in Fig. 5.3 abundance profiles in the ^{13}C pocket are plotted as functions of the initial number of protons, rather than of the position in mass. As shown in the figure, after all the mixed protons have been consumed, the abundance of ^{13}C presents a profile with a maximum corresponding to an initial mass fraction of protons of $\simeq 1\%$. This maximum value is defined by the initial amount of ^{12}C in the He intershell and the rates of the proton-capture reactions ^{12}C $+$ p and ^{13}C $+$ p, which produce and destroy ^{13}C, respectively [104, 175]. The neutron exposure profile is at its maximum in the layer corresponding to an initial mass fraction of protons $\simeq 0.3\%$. The difference between the location of the two maxima, that of the ^{13}C and that of the τ profile, is due to the presence of ^{14}N, which becomes important at the point where τ decreases after having reached its maximum. The ^{14}N nuclei have a relatively high neutron-capture cross section for the (n, p) reaction, thus capturing most of the free neutrons and decreasing the neutron flux.

Another interesting feature of the neutron flux in the ^{13}C pocket is its linear dependence on the initial amount of ^{12}C in the He intershell [175]. A higher ^{12}C abundance produces a higher abundance of ^{13}C, and thus a higher neutron exposure. Different initial amounts of ^{12}C in the

He intershell can be produced if overshoot is applied to the border of the convective pulse, resulting in the mixing of C-O core material, of roughly half carbon and half oxygen, into the intershell.

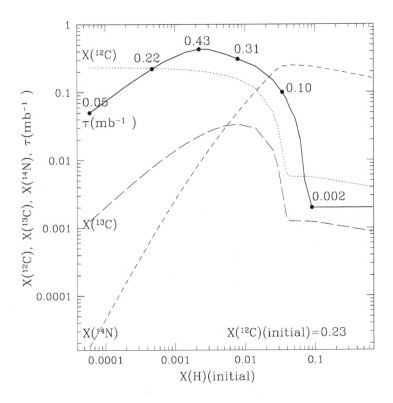

Fig. 5.3   Results of calculations performed for the nucleosynthesis in the ^{13}C pocket as a function of the initial number of protons mixed in the intershell. The mass fraction profiles of ^{12}C (dotted line), ^{13}C (long-dashed line) and ^{14}N (short-dashed line) are shown for the time when all the protons initially mixed have been captured. The neutron exposure, τ (solid line), is shown at the time when all the ^{13}C has been consumed by (α, n) reactions.

5.1.5   The neutron flux in the thermal pulse

In stellar models, the temperature, and hence the neutron flux in the convective pulse, varies as a function of the stellar mass and metallicity: it increases by either increasing the mass or decreasing the metallicity of the star (see e.g. [37]). It also depends on computational and physical choices, such as the details of the overshoot mechanism at the borders of the convective regions, when applied [175]. Moreover, it changes from one pulse to another within a given stellar model, as illustrated in Fig. 5.4, where the temperature and the neutron density in different pulses of the same stellar model are shown. The temperature, and hence the neutron density, increases with the pulse number, thus the effects of the activation of the ^{22}Ne neutron source are more prominent in the later rather than the earlier pulses.

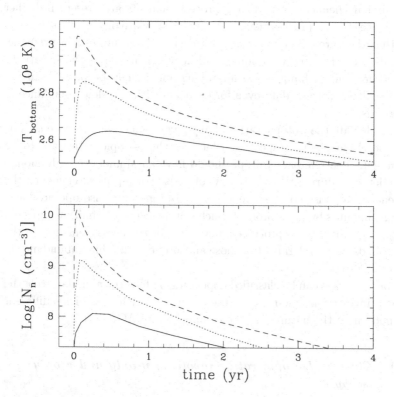

Fig. 5.4   Temperature (top panel) and neutron density (bottom panel) calculated for different thermal pulses of a solar metallicity 3 $M_\odot$ stellar model. The solid, dotted and dashed lines show the temperature and neutron density in the 4th, 9th and 18th (last) thermal pulses, respectively, that are followed by third dredge-up.

5.2   SiC grain data and the *s* process in AGB stars

Lugaro *et al.* [174] have performed a detailed comparison between AGB model predictions and SiC grain data for Zr, Sr, Mo and Ba. In this section, some of these authors' results will be presented and explained in detail. It is interesting to note that isotopic ratios of *all* heavy elements are somewhat affected by the *s* process and are expected to show a typical *s*-process signature in SiC grains. This is true also in the case of those elements that are actually mainly produced by the *r* process. The difference between *s*-process elements, such as Zr and Ba, and *r*-process elements, such as Eu and Pt, with regards to SiC grains is that, comparing elements with similar chemical properties, *s*-process elements are present in higher abundances than *r*-process elements, because they are produced in the star, and thus they are easier to measure. Hence, it is of interest, in principle, to analyse the isotopic variations of all heavy elements, as the sensitivity of measurement techniques are improving and it may be possible in the future to have isotopic data even for elements with very low abundance in SiC grains.

To calculate the isotopic ratios of any given element a reference isotope must be chosen, with respect to which all the isotopic ratios are to be calculated. The reference isotope should have an abundance high enough to be detected during the laboratory analysis. For *s*-process elements, the reference isotope is chosen to be an *s*-only isotope, or an isotope produced in large amounts by the *s* process. Such isotopes are by definition expected to be overabundant in *s*-processed material. For *r*-process elements, on the other hands, it is preferable to choose an isotope with a high abundance in the solar system.

Isotopic ratios can be classified depending on how the abundances of the isotopes involved are modified by the *s* process. Each class yields different information on the nature and the composition of AGB stars.

5.2.1   *Class I: Isotopic ratios involving p-only and r-only isotopes*

One class of isotopic ratios involves isotopes that are only produced by the *r* or the *p* process. As an example, Fig. 5.5 illustrates the case of the isotopic ratios involving three molybdenum isotopes: ^{92}Mo/^{96}Mo and ^{100}Mo/^{96}Mo, where ^{92}Mo is a *p*-only isotope, ^{96}Mo is the *s*-only reference isotope and ^{100}Mo is an *r*-only isotope (see Fig. 2.6). As in the case of the neon and

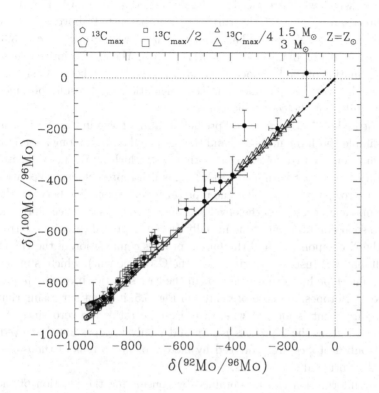

Fig. 5.5   Three-isotope plot for the ^{100}Mo/^{96}Mo and ^{92}Mo/^{96}Mo ratios measured in single SiC grains (black dots with 2σ error bars) and predicted by theoretical models of low-mass AGB stars of solar metallicity (small dots without error bars and open symbols connected by the solid lines) and masses 1.5 (small open symbols) and 3 $M_\odot$ (large open symbols).   Each symbol in the predictions corresponds to a third dredge-up episode. The small dots change into open symbols when the C/O ratio in the envelope is greater than unity, and hence SiC grains can form.   Isotopic ratios are given as δ values (see Sec. 4.1), so that $\delta=0$ corresponds to the solar ratio, which is the starting composition of the calculations.   For the ^{13}C pocket three choices are presented: one in which the efficiency of the s process in the ^{13}C-pocket region is dominated by the maximum value of neutron exposure achievable (Sec. 5.1.4), and another two cases in which the neutron exposure is scaled by a factor of 0.5 and 0.25, respectively.

silicon isotopic ratios presented in Figs. 4.2 and 4.8, it is most useful to represent isotopic ratios of heavy elements using three-isotope plots.   In this way it is possible to determine which different components and which degree of mixing between them have produced the observed compositions (see discussion in Sec. 4.4).   For heavy elements, the δ notation is usually employed, which was introduced for the silicon ratios in Sec. 4.1. Figure 5.5 is a three-isotope plot for the three Mo isotopes listed above. The Mo composition in single grains has been measured with RIMS (Sec. 3.3.3) [204], and model predictions are from [174]. Given the uncertainties related to the formation of the ^{13}C pocket (Sec. 5.1.2), the ^{13}C pocket is considered to be a relatively free parameter in the calculations and results obtained for three different choices are presented.

The trend of the observed and predicted ratios shown in Fig. 5.5 is relatively simple: both the measured and the predicted data lie along a straight line that connects two points: that with $\delta = 0$, which represents solar isotopic ratios, and that with $\delta = -1000^o/_{oo}$, which represents isotopic ratios equal to zero (pure ^{96}Mo). The line represents the mixing line between the two components present in the envelopes of carbon stars (see Sec. 4.4): the close-to-solar composition of the initially present material (usually referred to as the N component), and the pure s-process composition of the He intershell material (usually referred to as the G component), which is mixed into the envelope by third dredge-up. In the case of ratios involving p-only and r-only isotopes, as those presented in Fig. 5.5, it is not surprising that the G component is located where the isotopic ratios are zero since, by definition, during the s process p-only and r-only isotopes are destroyed, while s-only isotopes are produced by factors of up to tens of thousands above their original abundances.

Some information can be obtained by considering the location of the data points along the mixing line. They represent the degree of mixing between the two components, N and G. Any given mixture, in fact, falls at distances from the two components in inverse proportion to the number of atoms contributed by each component to the mixture. So, referring to the data points in Fig. 5.5, and taking $\delta = \delta(^{100}\text{Mo}/^{96}\text{Mo}) \simeq \delta(^{92}\text{Mo}/^{96}\text{Mo})$, one can calculate ratios such as:

$$\frac{\text{Mo}_G}{\text{Mo}_N} = \frac{\sqrt{2|\delta|^2}}{\sqrt{2|\delta + 1000|^2}},$$

where Mo_G and Mo_N are the numbers of atoms of Mo contributed by the G and the N components, respectively, and, the numerator and the denominator of the right term represent the distance of the mixture point from the N and the G component, respectively. On the other hand, the same ratio can be calculated using theoretical quantities (see Exercise 2.1) as:

$$\frac{\text{Mo}_G}{\text{Mo}_N} = OV^{\text{Mo}}_{s\,process} \times \frac{1}{DIL},$$

where

$$OV^{\text{Mo}}_{s\,process} = \frac{X^{\text{Mo}}_G}{X^{\text{Mo}}_N}$$

represents the s-process production factor of Mo in the He intershell with respect to the initial abundance, and

$$DIL = \frac{M_{envelope}}{M_{TDU}}$$

represents the dilution factor of the mass dredged-up from the He intershell into the envelope.

Thus data points in Fig. 5.5 in principle contain information on the efficiency of the s process in AGB stars since the $OV_{s\,process}$ depends on the efficiency of the s process in the ^{13}C pocket. For higher efficiencies, lower ^{100}Mo/^{96}Mo and ^{92}Mo/^{96}Mo are obtained as the s-only isotope ^{96}Mo is produced more and the p- and r-only isotopes ^{92}Mo and ^{100}Mo are destroyed more. Thus, model predictions in Fig. 5.5 get closer to the G component when the efficiency of the neutron flux, i.e. the amount of ^{13}C, is higher. However, the parameter DIL depends on the most uncertain AGB stellar structure features: the third dredge-up and the mass loss from the envelope. Hence it is not possible to derive conclusive constraints on any of these features from SiC data involving r- and p-only nuclei. Another problem is that possible pollution by material of solar composition during the measurements would draw the data points artificially closer to the N component.

Yet other information, relating to the initial composition of the star, can be derived. In the theoretical models the initial composition is assumed to be exactly equal to the composition of the sun. However, a weighted linear regression of the data points of Fig. 5.5 indicates that the N component extrapolated from the observations is somewhat enriched in ^{100}Mo, since at $\delta(^{92}$Mo/^{96}Mo)=0, $\delta(^{100}$Mo/^{96}Mo)=132 $\pm$ 41$^o/_{oo}$, suggesting a possible slight r-process enhancement in the initial composition of the parent stars of SiC grains.

5.2.2   Class II: Isotopic ratios involving isotopes in local equilibrium

Another class of isotopic ratios involves isotopes lying on the main s-process path and whose abundances are in local equilibrium $\langle\sigma_A\rangle N_A \simeq constant$ during the s process (Sec. 2.5.1).

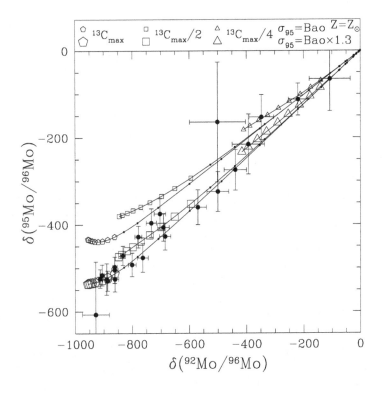

Fig. 5.6   Three-isotope plot of the ^{95}Mo/^{96}Mo and ^{92}Mo/^{96}Mo ratios. The symbols are as in Fig. 5.5, but only the 1.5 $M_\odot$ model is presented here and the larger open symbols represent the same cases as the smaller symbols, except that the value of the neutron-capture cross section of ^{95}Mo has been multiplied by a factor of 1.3 with respect to the recommended value.

Let us consider the isotopic ratio of ^{95}Mo/^{96}Mo as the example of an isotopic ratio involving nuclei in local equilibrium $\langle\sigma_A\rangle N_A \simeq constant$ (see Fig. 2.6). Isotopic data from single SiC grains and model predictions are shown in Fig. 5.6. Similar considerations as those presented in the previous section for the ^{100}Mo/^{96}Mo versus ^{92}Mo/^{96}Mo three-isotope plot apply here since both the data and the prediction points lie on the (almost, see discussion in Sec. 5.2.4) straight line that connect the initial envelope composition, N component, to the pure s-process composition from the He intershell, G component. However, in this plot the δ-value of the G component for the ^{95}Mo/^{96}Mo ratio, $\delta_G(^{95}$Mo/^{96}Mo), is not $-1000^o/_{oo}$, as in the case discussed in the previous section, because ^{95}Mo is produced by the s process, even if to a smaller extent than the s-only isotope ^{96}Mo. All model predictions in the plot tend to the same G component, for a given choice of the neutron-capture cross section of ^{95}Mo, $\langle \sigma_{95} \rangle$. This means that the ^{95}Mo/^{96}Mo ratio in the s-process material does not change significantly if we change either the amount of ^{13}C in the pocket, or the stellar mass (not shown in the plot). The ^{95}Mo/^{96}Mo ratio produced by the s process is, in fact, almost independent of the stellar model for which it is calculated because these two isotopes are in local equilibrium during the s process and hence their abundances follow the $\langle \sigma_A \rangle N_A \simeq constant$ rule (Sec. 2.5.1), which translates in this particular case to:

$$\frac{N_{95}}{N_{96}} \simeq \frac{\langle \sigma_{96} \rangle}{\langle \sigma_{95} \rangle},$$

where N_{95} and N_{96} are the s-process abundances of ^{95}Mo and ^{96}Mo, respectively, and $\langle \sigma_{95} \rangle$ and $\langle \sigma_{96} \rangle$ their neutron-capture cross sections. So, the ^{95}Mo/^{96}Mo isotopic ratio produced by the s process is largely determined by the inverse ratio of the neutron-capture cross sections of the two isotopes. Hence, precise information on the nuclear properties of the nuclei involved can be extracted, as demonstrated below.

In Fig. 5.6 the values predicted using as $\langle \sigma_{95} \rangle$ the recommended value from the most recent compilation of neutron-capture cross sections "Bao" [26], always lie above the data points. These model predictions in the plot tend to a G component of $\simeq -450^o/_{oo}$. The $\delta_G(^{95}$Mo/^{96}Mo) value derived from extrapolating the SiC data is instead $\simeq -550^o/_{oo}$, by about 100 lower than that extrapolated from the theoretical predictions. In the "Bao" compilation all data available on neutron-capture cross sections are reported and the recommended value for the ^{95}Mo neutron-capture cross section is taken from the latest available experiment. However, three previous experiments estimated $\langle \sigma_{95} \rangle$ to be approximately 30% to 40% higher than that recommended. As shown in Fig. 5.6, if predictions are computed using a 30% higher $\langle \sigma_{95} \rangle$ value, more ^{95}Mo is destroyed and the SiC grain data are better matched.

The possibility of predicting values of neutron-capture cross sections, hence constraining nuclear properties to a precision level that otherwise is only obtained by sophisticated nuclear experiments, demonstrates the unprecedented opportunities opened by measurements of heavy elements in SiC grains. This type of argument has been applied successfully to other isotopic ratios before, as in the case of Ba [100]. There are some puzzling cases, where calculations performed using the latest experimental neutron-capture cross sections do not match the SiC grain data, such as in the case of the $^{137}Ba/^{136}Ba$ ratio and of the neodymium isotopic ratios [98].

One final comment has to be made: in the discussion above the fact that neutron-capture rates can vary with the temperature at which the s process occurs, and thus the isotopic ratios of this class can actually depend on the s-process model considered, has not been taken into account. However, as observed in Sec. 2.5.1, typically, for neutron captures, the neutron-capture cross section $\sigma_A(v) \propto 1/v$ and thus the neutron-capture rates, which are proportional to $\langle \sigma_A(v)v \rangle$, are actually insensitive to temperature changes. It should be kept in mind, though, that for some special nuclei this may not hold, in which cases isotopic ratios depend on the temperature at which the s process occurs, even if the abundances of such nuclei are in local equilibrium $\langle \sigma_A \rangle N_A \simeq constant$.

5.2.3   Class III: Isotopic ratios involving isotopes with magic neutron numbers

As examples of this category of isotopic ratios let us consider two ratios of importance: $^{88}Sr/^{86}Sr$ and $^{138}Ba/^{136}Ba$. These were among the first ratios to be measured in SiC grains in bulk using TIMS (Sec. 3.3.3) [218, 219]. The first ratio involves ^{88}Sr, which has a magic number of neutrons (see Sec. 2.5) equal to 50, and the second ratio involves ^{138}Ba, which has a magic number of neutrons equal to 82. As discussed in Sec. 2.5.1, this type of nuclei represent a bottleneck for the neutron flux during the s process, since their neutron-capture cross sections are much smaller than those of the other heavy nuclei. How much of the neutron flux goes through these bottlenecks depends mostly on the value of the neutron exposure, as shown in Fig. 2.7: for different neutron exposures the steps in the $\langle \sigma_A \rangle N_A$ distribution at magic neutron numbers have different heights. It is clear from Fig. 2.7 that the production of ^{88}Sr and ^{138}Ba with respect to that of ^{86}Sr and ^{136}Ba, respectively, varies largely with the neutron exposure. For example, when the total neutron flux $\tau = 0.2$ mbarn^{-1}, both ^{88}Sr and ^{138}Ba have $\langle \sigma_A \rangle N_A$ values lower than those of ^{86}Sr and ^{136}Ba, respectively, while when $\tau = 0.9$

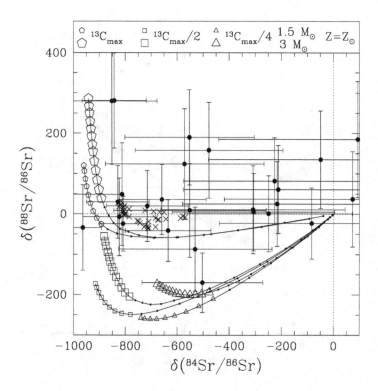

Fig. 5.7   Three-isotope plot for the ^{88}Sr/^{86}Sr and ^{84}Sr/^{86}Sr ratios. Symbols are as in Fig. 5.5, added are small crosses, which represent measurements of SiC grains in bulk.

mbarn^{-1} the reverse is true.

In Figs. 5.7 and 5.8 AGB model predictions and SiC grain data are compared. The Sr and Ba compositions have been measured in single SiC grains using RIMS [146, 205, 245] and in SiC grains in bulk using TIMS [224, 226]. Since the amount of ^{13}C in the pocket determines the neutron exposure, by varying it we obtain different predictions for the values of the isotopic ratios that involve neutron magic nuclei. In this situation model predictions define mixing lines that are well separated from each other, rather than very close, like in the cases discussed in the previous two sections.

Let us focus first on the information from measurements of SiC grains in bulk (represented by the small crosses in Figs. 5.7 and 5.8). When dealing with data from SiC grains in bulk it is usually preferable to derive the G

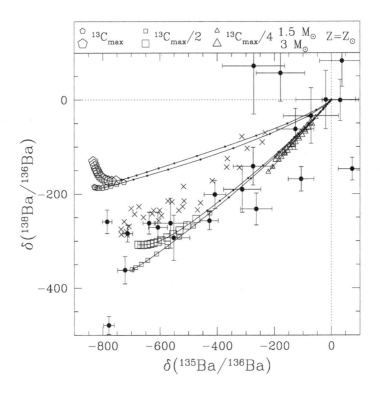

Fig. 5.8   Three-isotope plot for the ^{138}Ba/^{136}Ba and ^{135}Ba/^{136}Ba ratios. Symbols are as in Fig. 5.5, plus the small crosses, which represent measurements of SiC grains in bulk.

component to be compared with predictions for the pure s-process material. This is done to avoid problems related to the possibility of pollution of the sample with material of solar composition. In the case of the Sr isotopic ratios the G component is derived by imposing $\delta_G(^{84}\mathrm{Sr}/^{86}\mathrm{Sr}) = -1000^o/_{oo}$, since ^{84}Sr is a p-only nucleus. It is found that $\delta_G(^{88}\mathrm{Sr}/^{86}\mathrm{Sr}) \simeq 0$ for SiC grains in bulk. In the case of Ba, the G component is derived by imposing $\delta_G(^{130}\mathrm{Ba}/^{136}\mathrm{Ba}) = -1000^o/_{oo}$, since ^{130}Ba is a p-only nucleus. This corresponds to $\delta_G(^{135}\mathrm{Ba}/^{136}\mathrm{Ba}) = -850^o/_{oo}$ extrapolated in a ^{135}Ba/^{136}Ba versus ^{130}Ba/^{136}Ba three-isotope plot for SiC grains in bulk (shown in Fig. 5.11). The $\delta_G(^{135}\mathrm{Ba}/^{136}\mathrm{Ba})$ is a large negative number because ^{135}Ba is a nucleus with an odd number of nucleons, and has a relatively high neutron-capture cross section, so that it is largely destroyed during the s

process. Extrapolation in Fig. 5.8 gives $\delta_G(^{138}\text{Ba}/^{136}\text{Ba}) \simeq -350°/_{oo}$. It should be mentioned that small variations of the G components are derived from measurements of SiC grains of different sizes in bulk, with the $^{88}\text{Sr}/^{86}\text{Sr}$ and $^{138}\text{Ba}/^{136}\text{Ba}$ ratios decreasing with increasing the grains size. For both the $^{88}\text{Sr}/^{86}\text{Sr}$ and $^{138}\text{Ba}/^{136}\text{Ba}$ ratios the values of the G components are best reproduced by a case with a ^{13}C amount slightly higher than the case $^{13}\text{C}_{max}/2$. This means that, *on average* (since we are considering measurements if SiC in bulk, i.e. data from hundreds of thousands of grains) the SiC grain parent stars must have experienced a neutron exposure corresponding to such a case.

It was noticed in early studies [218] that such a value of the neutron exposure **does not** correspond to the value needed to reproduce the abundances of s-only isotopes in the solar system. The latter is instead represented by a larger value, close to that corresponding to the case with a maximum amount of ^{13}C in the pocket. This feature can be translated in an important piece of information on the average metallicity of the parent stars of SiC grains. In fact, one way to vary the neutron exposure in the ^{13}C pocket is by varying the metallicity of the star. The neutron exposure is directly proportional to the amount of the neutron source ^{13}C, and inversely proportional to the metallicity of the star, $\tau \propto ^{13}\text{C}/Z$. This is because for higher Z there is more material capturing neutrons, and hence the total number of free neutrons during the s process decreases, while the amount of ^{13}C is not a function of Z, because it is of primary origin being produced from the H and ^4He originally present in the star. Thus, the information yielded by isotopic ratios involving magic nuclei measured in bulk SiC samples, indicates that the SiC grain parent stars had, on average, a metallicity higher than that of the stars that produced the solar system distribution of s-process elements.

Since the average metallicity increases with the age of the Galaxy, we can conclude that SiC grains should have formed, on average, in stars that were born after those from which the bulk of the solar system material originated. This conclusion makes sense, if we consider that the bulk of the solar system material is the result of the operation of the Galactic chemical evolution, including the contributions of very old stars. On the other hand, SiC grains from very old stars would have spent a longer time in the interstellar medium than SiC grains from younger stars, and hence they would have preferentially been destroyed. As discussed in Sec. 4.8 a small population of SiC grains, the SiC-Z, are believed to have formed in stars of lower metallicities than the mainstream grain parent stars. Measurements of the isotopic composition of heavy elements in this type of SiC grains will be crucial in confirming the scenario described above.

With the advent of RIMS, we now have the possibility of considering also the information yielded by measurements of the Sr and Ba isotopic ratios in single SiC grains. For Sr and Ba, more grains show compositions close to the N component than in the case of Zr (see next section) and Mo. This can be explained by considering that SiC grains tend to have lower concentrations of Sr and Ba than of Zr and Mo, because Sr and Ba are more volatile than Zr and Mo [10, 155, 156]. Thus, Sr and Ba in SiC grains are more likely to be contaminated by solar system material than Zr and Mo.

With regards to the ^{88}Sr/^{86}Sr ratios, it is noticeable in Fig. 5.7 that a general trend of the lines of model predictions is to lie below the observed data. This may be considered as an indication that the initial composition (N component) for this ratio should be taken to be 10%–20% higher than solar. Another main feature of both the ^{88}Sr/^{86}Sr and ^{138}Ba/^{136}Ba ratios is that the composition of single grains is not concentrated around the composition of SiC grains in bulk, but shows a large spread. For example, $\delta(^{88}\text{Sr}/^{86}\text{Sr})$ spreads from -200 to $+300^o/_{oo}$, from which one can deduce that a spread in the efficiency of the neutron flux in the ^{13}C pocket must be present for stars of a given mass and metallicity. The same conclusion is reached when model predictions are compared with spectroscopic observations of heavy s-process elements in stars of different metallicities [48].

To explain the existence of a spread of neutron efficiencies in the ^{13}C pocket for a given stellar model one can use stellar rotation, as proposed by Herwig *et al.* [118]. All stars rotate and, when rotation is taken into account in stellar evolution models, it is found that the region where the ^{13}C pocket is supposed to form in the current s-process model is not in stationary conditions during the interpulse period. Instead, material is continuously mixed due to rotationally-induced shear. Shear mixing is produced by the fact that the rotational velocities of the core and the envelope are different, since the compact degenerate core rotates faster than the less dense extended envelope. The border between the core and the envelope, which experiences the discontinuity in rotational velocity and hence is the place where the shear mixing occurs, it is also the place where the ^{13}C pocket is located. Thus, the evolution in time of the abundances of ^{13}C and ^{14}N in each layer of the ^{13}C pocket, whose initial profiles are shown in Fig. 5.3, when rotation is included, do not depend only on the effect of nuclear burning, but also on the effect of mixing in the region. Under radiative conditions, ^{14}N has a high abundance only in the layers of the pocket that have seen a mass fraction higher than $\simeq 1\%$ of protons diffused from the envelope. In the presence of rotation, this ^{14}N is mixed down into ^{13}C-rich layers. Because ^{14}N is a strong neutron absorber, it follows that the neutron flux in the pocket is modulated by the shear mixing. The final neutron exposure in each layer of the pocket is thus decreased with respect to the values shown in Fig. 5.3.

The problem is that current stellar models with rotation predict that the rotational mixing in the ^{13}C pocket completely inhibits the occurrence of the s process! In order to produce a rotating stellar model in which the s process is not absent, but only modulated by shear mixing, other effects should be taken into account. For example, the inclusion of magnetic fields in the computation could help to decrease the amount of shear mixing in the ^{13}C pocket. Magnetic fields could enhance the coupling of the core to the envelope of the star, thus producing a smaller difference between the rotational velocities of the core and the envelope. This would result in a weaker shear-mixing effect at the border between the core and the envelope.

Measurements of isotopic ratios involving magic nuclei in single presolar grains can set strong constraints on the way the ^{13}C neutron source operates in AGB stars, and on processes such as stellar rotation and the presence of magnetic fields.

5.2.4 Class IV: Isotopic ratios involving isotopes depending on branchings

Let us now consider two Zr isotopic ratios: ^{91}Zr/^{94}Zr and ^{96}Zr/^{94}Zr. A three-isotope plot is shown in Fig. 5.9, with single SiC grain data and model predictions. The Zr composition in single SiC grains has been measured using RIMS [203]. Note that, before RIMS, measurements of the isotopic composition of Zr and Mo in SiC grains could not be performed, since it was impossible to discriminate between isotopes of the same atomic mass, such as ^{96}Zr and ^{96}Mo. The data points lie closer to the G than to the N component, similarly to the Mo isotopic ratios of Figs. 5.5 and 5.6. This is consistent with the fact that Zr and Mo are present in SiC grains in larger amounts than Sr and Ba, hence contamination by solar system material has less effect.

The ratio ^{91}Zr/^{94}Zr is a good example to summarise the types of information discussed in the previous two sections. Zirconium-91 is very close to having a magic number of neutrons, with $N=51$ (Fig. 2.6). Hence the isotopic ratio $^{91}Zr/^{94}Zr$ behaves similarly to the ratios of class III (Sec. 5.2.3). In particular, it varies over a large range of values when changing the ^{13}C amount, and hence the neutron exposure in the ^{13}C pocket. Single grain data show a spread of $\delta(^{91}Zr/^{94}Zr)$ from -300 to almost $+100^o/_{oo}$. This is in agreement with the conclusion drawn from the Sr and Ba compositions in single grains, that a spread in the neutron exposure in the ^{13}C pocket is needed to cover the range of data.

Model predictions, though, are systematically higher than SiC data, for the $^{91}Zr/^{94}Zr$ ratio. The neutron-capture cross sections of ^{91}Zr has an uncertainty of 13% [26] and a 10% higher cross section for this isotope would lead to δ-values about $100^o/_{oo}$ lower than those shown in Fig. 5.9, and hence a better match with the measurements. As in the case discussed in Sec. 5.2.2, data of Zr isotopic ratios in SiC grains give us the possibility of high-precision testing of the nuclear properties of the Zr isotopes, therefore also stimulating new laboratory measurements, since available data date back to the 1970s.

The other ratio shown in Fig. 5.9 is that involving ^{96}Zr. This is a particularly interesting isotope: it would in principle be considered as an r-only isotope, however, it can be produced during the s process if the branching point at ^{95}Zr is activated, as shown in Fig. 2.8. The final abundance of ^{96}Zr largely depends on the maximum neutron density achieved during the s process. Since high neutron densities are reached during the neutron flux in the convective pulse (see Sec. 5.1.5), the abundance of ^{96}Zr can be used to test the properties of this neutron flux. The lines of model predictions start from solar and then progressively move towards more and more negative $\delta(^{96}Zr/^{94}Zr)$ values, until a minimum is reached and the prediction turns back towards higher $\delta(^{96}Zr/^{94}Zr)$. This behaviour indicates that the G component for this ratio is changing after each thermal pulse! This change is due to the fact that the temperature in the convective pulse increases with the pulse number, as shown in Fig. 5.4. Hence, the composition of the material dredged-up after each pulse is unique. In particular, more ^{96}Zr is produced during the late higher-temperature pulses, which explains the turning in the predicted evolution lines of Fig. 5.9.

A similar curve is also noticeable in the predictions shown in the three-isotope plots presented in the previous two sections and can now be interpreted. In the case of $\delta(^{95}Mo/^{96}Mo)$ there is a very small curving toward higher ratios. This is due to the different way in which, toward the end of the neutron flux in each convective pulse, the branching at ^{95}Zr toward

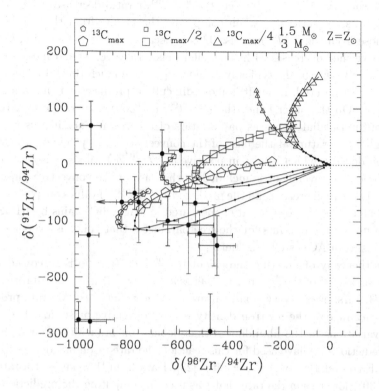

Fig. 5.9   Three-isotope plot of the ^{91}Zr/^{94}Zr and ^{96}Zr/^{94}Zr ratios. Symbols are as in Fig. 5.5.

^{96}Zr progressively closes as the neutron density decreases. Also in the case of $\delta(^{88}$Sr/^{86}Sr$)$ and of $\delta(^{138}$Ba/^{136}Ba$)$ the predicted evolution as a function of the pulse number shows a turning toward high values. The abundances of magic nuclei always tend to accumulate because of their small neutron-capture cross sections, and, in fact, their abundances slightly grow in the convective pulses. Since the small neutron exposure produced in the pulses increases, together with the peak neutron density, with the pulse number, higher abundances of the magic nuclei are produced in the later pulses.

Single SiC grains show a spread in $\delta(^{96}$Zr/^{94}Zr$)$ from about $-500^o/_{oo}$ to $-1000^o/_{oo}$. The grains with composition close to $-1000^o/_{oo}$ are quite extreme since this value of δ represents a composition in which no ^{96}Zr is produced during the s-process. Such grains are not currently matched by the model predictions. One possibility is that their parent stars had mass lower than 1.5 $M_{\odot}$. In fact, the ^{96}Zr/^{94}Zr ratio decreases with the stellar mass, as noticeable when comparing the results from the 1.5 $M_{\odot}$ and those from the 3 $M_{\odot}$ models. This is because the temperature in the convective shell decreases with the stellar mass. However, stellar models of mass lower than 1.5 $M_{\odot}$ typically do not experience much third dredge-up of intershell material, hence it is more difficult to produce carbon stars in this case. On the other hand, the low $\delta(^{96}$Zr/^{94}Zr$)$ values rule out AGB stars of intermediate mass as parent stars of the grains since in this case the high temperature at the base of the convective pulses produces a high neutron density, which results in positive $\delta(^{96}$Zr/^{94}Zr$)$ values. Also, when time-dependent overshoot is included at the base of the convective pulse, the temperature increases and the predicted $\delta(^{96}$Zr/^{94}Zr$)$ values are positive [175]. Thus, the Zr isotopic ratios measured in single SiC grains by RIMS can be used as constraints for the free parameter that controls overshoot mechanisms in AGB stellar models.

Another way of lowering the calculated ^{96}Zr/^{94}Zr ratio is to consider the uncertainties of the ^{22}Ne$(\alpha, n)^{25}$Mg reaction rate. During the s-process in AGB stars, even a very small variation of the ^{22}Ne$(\alpha, n)^{25}$Mg rate produces a change of the neutron density in the convective pulse, and hence some variations in the abundances that depend on branching points. In our predictions we have used for this reaction the rate recommended in the NACRE compilation of reaction rates [23]. Data from [145] suggest a rate about 30% lower than the rate that was used in computing the predictions shown in Fig. 5.9 [23, 154]. A rate 30% lower results in a predicted ^{96}Zr/^{94}Zr ratio about 4% lower, and the lowest $\delta(^{96}$Zr/^{94}Zr$)$ value obtained, for the case of 1.5 $M_{\odot}$ and the maximum amount of ^{13}C, moves from $-830^o/_{oo}$ shown in Fig. 5.9 down to $-870^o/_{oo}$. In any case the fact that the Zr composition of presolar SiC grains shows negative $\delta(^{96}$Zr/^{94}Zr$)$ values rules out a rate for the ^{22}Ne$(\alpha, n)^{25}$Mg reaction much larger than what we have used, such as the upper limit for the rate proposed in the NACRE compilation.

Another point to consider is that the neutron-capture cross section of the nucleus that governs the branching, ^{95}Zr, is very uncertain. For unstable isotopes, it is much more difficult to perform experiments to measure their neutron-capture cross sections than for stable isotopes. Hence, only theoretical estimates are usually available, and they have typical uncertainties of 50%. For example, for the neutron-capture cross section of ^{95}Zr there are different estimates ranging from 23 mbarn [149] to 126 mbarn [229] (values given at a temperature of 30 keV). In the predictions presented in Fig. 5.9 we have used the recommended value of 79 mbarn [26]. If we use the lowest available estimate the $^{96}Zr/^{94}Zr$ ratio decreases by about 5% so that the lowest $\delta(^{96}Zr/^{94}Zr)$ we obtain is $-880^o/_{oo}$. Using instead the highest estimate available, the lowest $\delta(^{96}Zr/^{94}Zr)$ value is $-780^o/_{oo}$.

There are many isotopes that are involved in branchings and, as in the example discussed above, their abundances in SiC grains hold a great deal of information on stellar structure and nuclear properties. More examples are the cases of ^{86}Kr (see below), ^{134}Ba, ^{137}Ba, and ^{148}Nd [98, 174].

5.2.5 Class V: Isotopic ratios involving isotopes produced by radioactive decay

As in the case of ^{26}Al (Sec. 4.5), there are some heavy long-lived radioactive nuclei that could have been incorporated in SiC grains. The best example is ^{99}Tc, which decays into ^{99}Ru with a half-life of 2.1×10^5 years. As mentioned at the beginning of this chapter, the presence of Tc in AGB stars is convincing evidence that the s process occurs in these stars. The proof of the initial presence in SiC grains of ^{99}Tc, that has now all decayed, is shown by the Ru composition measured in single SiC grains with RIMS [246]. The data show the typical signature of the s process, with all the isotopic ratios, calculated using the s-only isotope ^{100}Ru as the reference isotope, lower than solar, in particular with the p-only $^{96,98}Ru$ and the r-only ^{104}Ru being strongly depleted. The production of ^{99}Ru during the s process is due to the decay of ^{99}Tc, which is on the main s-process path (Fig. 2.6). The decay of ^{99}Tc into ^{99}Ru occurs both in the He intershell and in the envelope. As illustrated in Fig. 5.10, SiC grain data for the $^{99}Ru/^{100}Ru$ ratio are matched, only if it is assumed that Tc is incorporated in the grains together with Ru, so that the abundance at mass 99 in the grains reflects both the contributions of the ^{99}Ru and the ^{99}Tc originally present in the stellar envelope. This is a reasonable assumption, as Tc is a refractory element, as is Ru. The detection of the initial presence of ^{99}Tc in SiC grains is further proof of their origin in AGB stars.

A different conclusion is reached when considering the possible contribution to the abundance of ^{135}Ba from the presence of the long-living ^{135}Cs (with half-life of 2 million years) in single SiC grains [174]. Cesium does not appear to have condensed in the grains. In fact, if it is assumed that Cs is also incorporated in SiC grains, then the predicted $^{135}Ba/^{136}Ba$ ratios are higher than the measured values, as illustrated in Fig. 5.11. This is consistent with the fact that Cs is a highly volatile element and thus it is not expected to have condensed into SiC grains.

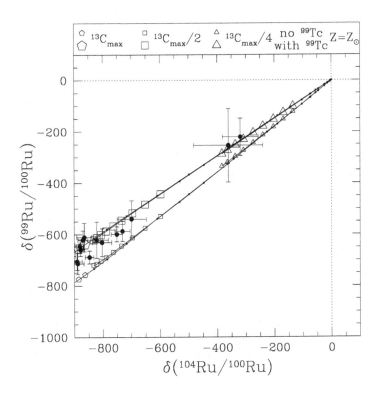

Fig. 5.10   Three-isotope plot of the ^{104}Ru/^{100}Ru and ^{99}Ru/^{100}Ru ratios. The symbols are as in Fig. 5.5, but only the 1.5 $M_\odot$ model is presented. The larger open symbols represent the same cases as the smaller open symbols, except that the abundance of ^{99}Tc in the envelope of the star is or is not added to the abundance of ^{99}Ru, respectively.

5.3   The heavy noble gases: Kr and Xe

The heavy noble gases Kr and Xe are discussed separately from the other heavy elements because, unlike the mostly refractory elements discussed above, noble gases are extremely volatile and are believed to have been implanted into SiC grains after having been ionised [168]. The available data are for SiC grains in bulk [168], and are shown in Figs. 5.12 and 5.13.

In Fig. 5.12, the Xe-S component (also shown in Fig. 1.3) is plotted and compared to AGB model predictions. The G component is considered, since data are for SiC grains in bulk and contamination with solar material could have happened. The G components for SiC grains is extrapolated

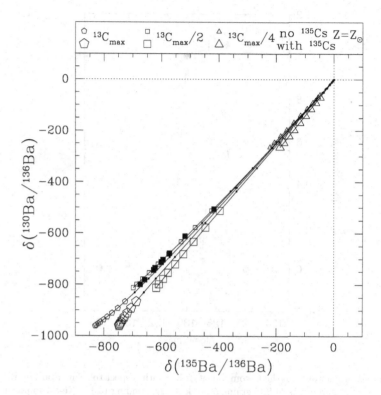

Fig. 5.11   Three-isotope plot of the ^{135}Ba/^{136}Ba and ^{130}Ba/^{136}Ba ratios. The symbols are as in Fig. 5.5, but measurements are only for SiC grains in bulk (black squares) and only the 1.5 $M_\odot$ model is presented. The larger open symbols represent the same cases as the smaller open symbols, except that the abundance of ^{135}Cs in the envelope of the star is or is not added to the abundance of ^{135}Ba, respectively.

by setting ^{136}Xe/^{130}Xe $= 0$, since ^{136}Xe is an r-only isotope. This is com-pared with model predictions for the pure s-process component, i.e. for material in the He intershell. To take into account the fact that the He intershell composition changes after each pulse and also that the amount of third dredge-up varies, predicted values are calculated as the average of the composition after each pulse, provided that C/O > 1 in the envelope at such time, weighted by the mass dredged-up after each pulse. Moreover, since we compare predictions with data from SiC grains in bulk another average is performed on the possible choices of the amount of ^{13}C in the pocket, for stellar masses of 1.5 and 3 $M_\odot$, and for metallicities solar and

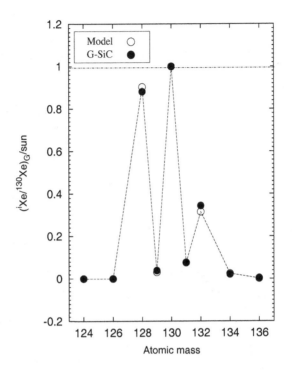

Fig. 5.12 Isotopic ratios of the G component of Xe, with respect to solar ratios, extrapolated from measurements of SiC grains in bulk (corresponding to the Xe-S component shown in Fig. 1.3, error bars are within the symbols) and AGB model predictions for pure s-process intershell material, calculated as explained in the text (courtesy Marco Pignatari).

half of solar. Compositions from all these models are weighted by the total abundance of Xe present in the stellar winds for each case.

Among the Xe isotopic ratios, some belong to the ratios of class I discussed in Sec. 5.2.1, which involve p-only (^{124}Xe and ^{126}Xe) and r-only isotopes (^{134}Xe and ^{136}Xe). All the remaining ratios involve isotopes on the main s-process path, and far from magic neutron numbers (class II, Sec. 5.2.2). Hence, the data yield mostly information on the neutron-capture cross section of the Xe isotopes with A from 128 to 132. Model predictions represent a good match with the data, in particular since recent measurements of the neutron-capture cross sections of the ^{128,129,130}Xe isotopes have reduced their uncertainties by an order of magnitude [231].

By applying a simple ion implantation model to the concentration of Xe in SiC grains of different sizes, it is possible to derive information on the energy of the stellar winds in which these gases were ionised and implanted into the grains [286]. The Xe concentration as a function of the grain size can be matched if the implantation of Xe has occurred in relatively slow stellar winds of 10–30 km/s, typical of the AGB phase. The Xe composition, as that of the other heavy elements discussed above, shows the mixing of the N and G components in the envelope of AGB stars.

When the implantation model is applied to the concentration of the light noble gases, He, Ne and Ar, in SiC grains of different size (whose isotopic composition was discussed in Sec. 4.4) it appears, instead, that these gases were implanted in SiC grains in higher-energy stellar winds, with velocities of a few thousand km/s. These conditions can be found in post-AGB stars, when the last fraction of the envelope of the star is ejected. At this point of the AGB evolution, the H-rich envelope has a very small mass and hence it is easier to achieve small dilution of the He intershell material. This is in agreement with the fact that the isotopic compositions of He, Ne and Ar in SiC grains are very close to the composition of the G component, i.e. they represent almost pure He intershell material. Note that the ionisation potential is higher for these elements than for Xe, hence higher energies are needed to ionise them.

A more complex situation arises for Kr, whose ionisation potential is half way between those of He, Ne and Ar and that of Xe. The G component of the Kr isotopic composition measured in SiC grains in bulk is shown in Fig. 5.13 and compared to AGB model predictions computed in the same way as for Xe. Variations of the isotopic composition are interestingly correlated with the size of the grains: in particular the ^{86}Kr/^{82}Kr ratios span a large range of values increasing with the grain size, in opposite direction to the small range spanned by the ^{88}Sr/^{86}Sr and ^{138}Ba/^{136}Ba ratios measured in SiC grains in bulk (Sec. 5.2.3).

The concentration of Kr in SiC grains as a function of the grain size can be explained by the implantation model if two Kr components are present: one that was implanted in low-energy stellar winds, similarly to Xe, and is responsible for the low ^{86}Kr/^{82}Kr ratios in the small grains, and another that was implanted in high-energy stellar winds, similarly to He, Ne and Ar, too energetic to be captured by the small grains and thus is responsible for the high ^{86}Kr/^{82}Kr ratios in the larger grains. Further work on the nucleosynthesis in AGB and post-AGB stars is needed, in order to explain the features of the isotopic patterns of Kr in SiC grains within the complex scenario described above.

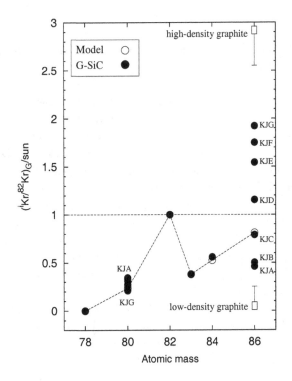

Fig. 5.13 Isotopic ratios for the G component of Kr, with respect to solar ratios, extrapolated from measurements of SiC grains in bulk (the size of the grains are indicated for ^{80}Kr/^{82}Kr and ^{86}Kr/^{82}Kr, see Fig. 3.1, error bars are within the symbols) and AGB model predictions for pure s-process intershell material, calculated as explained in the text (courtesy Marco Pignatari and Carlo Fazio). The G components for the ^{86}Kr/^{82}Kr ratio extrapolated from measurements of presolar graphite grains in bulk of different densities (see Sec. 6.2) are also shown for comparison.

Not only the ^{80}Kr/^{82}Kr and ^{86}Kr/^{82}Kr ratios are interesting because they both vary, slightly or strongly, respectively, with the grain size, but also because their production is dependent on the features of the stellar model. The ^{80}Kr/^{82}Kr ratio depends on the temperature at which the s process occurs, because the production of ^{80}Kr is affected by a branching point at ^{79}Se, whose β^--decay rate is strongly dependent on temperature. The ^{86}Kr/^{82}Kr ratio is strongly dependent on the neutron density at which the s process occurs because the production of ^{86}Kr is affected by the branching point at ^{85}Kr (see Fig. 5.14). Krypton-85 has a ground state with a

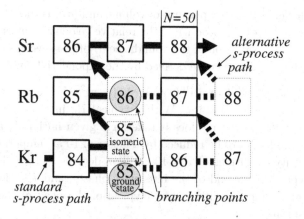

Fig. 5.14 Section of the nuclide chart showing the branching points at the ground state of ^{85}Kr (half-life of 11 years) and at ^{86}Rb (half-life of 19 days). When either of the branchings is open, part of the flux goes through ^{86}Kr and ^{87}Rb (which have magic number of neutrons =50), skipping ^{86}Sr and ^{87}Sr.

half-life of 11 years, much longer than that of the isomeric state[2], and hence can easily capture neutrons before decaying. For example, about 20% of the neutron flux goes through ^{86}Kr for neutron densities as low as 4×10^7 cm^{-3}. Moreover, ^{86}Kr itself has a magic number of neutrons ($N=50$) so its neutron-capture cross section is relatively low and, once produced, ^{86}Kr tends to accumulate. One problem is that the neutron-capture cross section of the unstable ^{85}Kr is uncertain by almost a factor of two so that current AGB models can span the whole range of the ^{86}Kr/^{82}Kr ratio observed in SiC grains of different sizes just by considering the range of uncertainty in the neutron-capture cross section of ^{85}Kr! This fact precludes any confident conclusions, however, once this uncertainty will be reduced, future developments could even aim at relating nucleosynthesis conditions in AGB and post-AGB stars to the energy of the related stellar winds, through the Kr composition of presolar SiC grains.

5.4 Exercises

(1) The lifetime of a nuclear species i against capturing a particle of species j can be calculated as $\tau = 1/\lambda = 1/(n_j \langle \sigma v \rangle)$ s^{-1}, where n_j is the num-

---

[2]Excited *metastable* state, i.e. with a longer lifetime with respect to ordinary excited states, before decaying into the ground state (of 4.5 hours in the case of ^{85}Kr).

ber of nuclei of species j per unit of volume and $\langle \sigma v \rangle$ is the Maxwellian-averaged value of the product of the relative velocity v times the cross section σ. The variation of the abundance n_i of the species i at a given time t, dn_i/dt, can then be derived by the exponential law $e^{-t/\tau}$.

a) Consider the ^{13}C$(\alpha, n)^{16}$O and ^{22}Ne$(\alpha, n)^{25}$Mg reaction rates from Fig. 5.1 in AGB stars. With a typical He intershell abundance $X_4 = 0.7$, how long does it take for a given initial abundance of ^{13}C and ^{22}Ne to be reduced by a factor of 10 at temperatures and density conditions of $T = 100$ million degrees and $\rho = 10^4$ g/cm^3 (typical of the end of the interpulse phase), and $T = 300$ million degrees and $\rho = 10^3$ g/cm^3 (typical of the peak of the thermal instability)?

b) During each interpulse-pulse cycle the temperature in the He intershell for a typical solar metallicity AGB star of mass $\sim 2\ M_\odot$ stays above 100 million degrees for ~ 1000 years at the end of the interpulse period, and above 300 million degrees for ~ 0.5 year during the thermal pulse. Which of the two neutron source reactions described above is more likely to contribute neutrons to the s process in this star?

(2) Consider a single SiC grain with composition $\delta(^{92}$Mo$/^{96}$Mo$) = \delta(^{100}$Mo$/^{96}$Mo$) = -700^o/_{oo}$. Calculate what fraction of Mo atoms have contributed to the total amount of Mo in such a grain by the N and by the G components. Remember that in a three-isotope plot the number of nuclei contributed by the N and G component to the given mixture is inversely proportional to the distance of the mixture point from the two components.

(3) Consider the isotopic ratio of the two s-only isotopes ^{128}Xe$/^{130}$Xe which are in local equilibrium $\langle \sigma_A \rangle N_A \simeq constant$ during the s process (see Fig. B.1). In the compilation of Ref. [26] the neutron-capture cross sections for these two isotopes are $\langle \sigma_{128} \rangle = 248$ mbarn and $\langle \sigma_{130} \rangle = 141$ mbarn, at a temperature of 30 keV. The G component of SiC grains in bulk shows $(^{128}$Xe$/^{130}$Xe$)_{SiC} = 0.44$. Is this value reproduced in pure s-process material?
Reifarth *et al.* [231] measured $\langle \sigma_{128} \rangle = 262.5$ mbarn and $\langle \sigma_{130} \rangle = 132.0$ mbarn (given for a temperature of 30 keV). Is the SiC grain value more closely reproduced using these latest data for neutron-capture cross sections? Compare with the results presented in Fig. 5.12 (it is useful to also consider Fig. B.1).

(4) Derive the $\langle \sigma_A \rangle N_A$ values plotted in Fig. 2.7 for ^{86}Sr and ^{88}Sr. Knowing that $\langle \sigma_{86} \rangle / \langle \sigma_{88} \rangle \simeq 10$ and that ^{88}Sr$/^{86}$Sr $= 8.34$ in the solar system, calculate $\delta_G(^{88}$Sr$/^{86}$Sr$)$, i.e. the composition of pure s-process material, for the four cases of neutron exposure presented in Fig. 2.7. Where would these values plot in Fig. 5.7?

(5) Knowing that $\langle \sigma \rangle v_{th}$ for ^{85}Kr is 1.4×10^{-17} cm^3 s^{-1} at 23 keV and that its half-life is 11 yr, demonstrate that there is 20% probability that the s-process path branches towards ^{86}Kr when the neutron density is $\simeq 4 \times 10^7$ cm^{-3}.

Chapter 6

Diamond, Graphite and Oxide Grains

Presolar diamond, graphite and oxide grains have been less studied and analysed than silicon carbide grains. This is mostly because it is practically more troublesome to extract, identify and analyse these types of grains. However, current technological advances in the procedures of the laboratory analysis of the grains, are leading the way to bright future prospects with regards to the extraction of more information. In this chapter we review the available data, their theoretical interpretation and future possibilities in the study of these types of presolar grains.

6.1 Diamond

The feature of presolar diamond that most stands out is the composition of the noble gas Xe carried by the grains: the Xe-HL component that was one of the first signs of the presence of presolar material in meteorites (see Sec. 1.2 and Fig. 1.3). Although among the first to be discovered, the Xe-HL isotopic pattern is still the least understood. Most of the effort has so far been concentrated on explaining the Xe-H component, i.e. the composition of the heavy Xe isotopes ^{134}Xe and ^{136}Xe. Since these isotopes are r-only nuclei, the most natural explanation is to relate their enhancements to the occurrence of the r process. However, the main problem is that the ^{136}Xe/^{134}Xe ratio is different from the solar ratio (Fig. 1.3). Hence, the r process that produced the Xe-H composition cannot be the same as the r process that produced the solar system r-process abundance distribution.

A possible scenario for reproducing the Xe-H composition involves a short and intense neutron burst in the He-rich zone of SNII [63] (see Sec. 2.5.2). This process is quite unlike the classical r process because the neutron burst is not as strong, but rather presents intermediate features

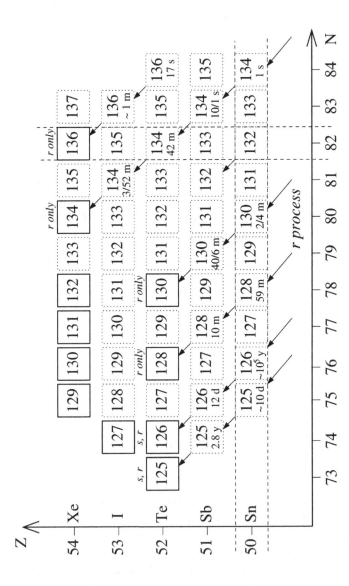

Fig. 6.1   Section of the nuclide chart (similar to Fig. 2.6) including some isotopes of elements from Sn to Xe. Long-dashed lines highlight the magic numbers of protons (50) and and neutrons (82). The *r*-process paths of production, through β^--decays, of the *r*-only isotopes ^{128,130}Te, as well as of ^{125,126}Xe, which are of mixed *s*- and *r*-origin, are indicated by the arrows. The half-lives of the unstable isotopes on such paths are also indicated (where "s" is for seconds, "m" for minutes, "d" for days and "y" for years). In some cases two half-lives are indicated corresponding to different isomeric states of the nucleus.

between those of the the r and the s processes. By means of a quantitative model, Howard *et al.* [130] calculated which possible neutron irradiation histories could have produced the observed Xe-H composition, and made predictions regarding the composition of other elements. It is noteworthy that this same model can also be applied to explain the composition of Zr and Mo in single SiC grains of type X (Sec. 4.8) and the Zr composition of certain single graphite grains (Sec. 6.2), which is consistent with a common SNII origin for these three types of presolar material.

Another model to explain the Xe-H component in diamond was proposed by Ott [217]. In this model the classical r process is followed by the separation of xenon from iodine and tellurium. During the r process, the abundances of ^{134}Xe and ^{136}Xe are produced after the neutron flux is extinguished and neutron-rich nuclei with A=134 and 136 undergo β^- decay toward their corresponding stable isobars. These production paths are illustrated by the arrows in Fig. 6.1. In the case of ^{136}Xe, the β^--decay half-lives of all its unstable precursors are of the order of 1 minute or less, hence its abundance is quickly established. In contrast, in the case of ^{134}Xe, its precursors ^{134}I and ^{134}Te have β^--decay half-lives of the order of several tens of minutes. Thus, if within approximately two hours Xe is separated from I and Te, the isotopic pattern of Xe-H can be reproduced. Ott [217] proposed different ways by which this separation might occur: fast condensation of Te and I, which are more refractory than Xe; loss of ^{134}Xe from the grains by recoil during the β^--decay of its precursors previously trapped in the grains; and separation of ionised from neutral atoms, maybe by means of magnetic fields, given that Xe atoms are more difficult to ionise.

The separation model was subsequently shown to be successful also in explaining the composition of tellurium in diamonds [237]. The two Te isotopes of mixed r and s origin, ^{125}Te and ^{126}Te, show no excesses in diamonds, with respect to the solar composition. An excess is predicted by the neutron-burst model, but not by the separation model, since the r-process precursors of ^{125}Te and ^{126}Te have very long half-lives against β^--decay, of the order of years (Fig. 6.1). On the other hand, diamonds show a very slight enhancement of the r-only isotope ^{128}Te with respect to the other r-only isotope ^{130}Te, while the separation model predicts the reverse: a slight enhancement of ^{130}Te with respect to ^{128}Te. This is because the production of ^{128}Te is controlled by the decay half-life of ^{128}Sn (59 minutes), which is longer than the decay half-life of any of the ^{130}Te precursors. Given that the neutron-burst and the rapid-separation models both have some advantages and problems, a hybrid model may be considered, in which isotopic compositions produced by a neutron burst are followed by rapid separation [179].

The Xe-L component, in which the ratio of the two p-only isotopes ^{124}Xe and ^{126}Xe is about two times higher than solar, can be qualitatively explained within the separation scenario. In fact, the p-process production of ^{126}Xe occurs via β^+ decay of ^{126}Ba, which has a longer half-life (100 minutes) than ^{124}Ba (12 minutes), the p-process precursor of ^{126}Xe. However, the ^{128}Xe$/^{126}$Xe ratio predicted by models of the p process in SNII is much higher than that observed in Xe-HL [230]. This problem prevents us from drawing any confident conclusions. Pioneering work on the Xe-L component was also presented in [119, 120].

An alternative scenario was proposed by Jørgensen [148], where diamonds originating in red giant stars in binary systems are implanted with a Xe-HL component produced by the SNIa explosion of the white dwarf companion of the giant. Other proposed stellar sources for diamond grains include C-rich giant stars [22] and Wolf-Rayet stars [25].

Being the grains too small, the analysis of the isotopic composition of individual diamond grains, even for the most abundant element, carbon, is outside the current laboratory possibilities, but may be achieved in the future. Information on individual presolar diamond would represent a better and perhaps only way to understand where these grains actually formed and which fraction of them are of truly presolar origin.

6.2   Graphite

Presolar graphite grains are classified by their density and, interestingly, the carbon isotopic ratios and the noble gas compositions vary with the density [9, 13, 123]. The ^{12}C$/^{13}$C ratio varies from $\simeq 2$ to $\simeq 10,000$ providing good evidence for a stellar origin. Grains with ^{12}C$/^{13}$C higher than solar (89) are more abundant among grains with high density. On the other hand, the nitrogen isotopic ratios are mostly close to the terrestrial value (272), perhaps indicating that at least part of the nitrogen in graphite equilibrated with air. With regard to the noble gases, presolar graphite grains are the carriers of the Ne-E(L) component [13], i.e. monoisotopic ^{22}Ne most likely resulting from the radioactive decay of ^{22}Na produced in novæ and supernovæ. In addition, graphite grains also carry a milder anomaly similar to that of Ne from the He intershell of Asymptotic Giant Branch stars (as in the case of mainstream SiC grains, Sec. 4.4). The Kr composition also appears intriguing (Fig. 5.13), showing the signature of the s process occurring both at low and high densities. Thus, a variety of stellar sources have to be considered for the origin of graphite grains, from very massive stars, to supernovæ, novæ and AGB stars. This multiplicity of sources is confirmed by recent measurements in single high-density graphite grains of He and Ne [202] and the heavy elements Mo and Zr [206]. Graphite grains of high density, in particular, are believed to have originated from red giant and Asymptotic Giant Branch stars, however, it is not clear why there are so few of them with respect to presolar SiC grains. Overall, the dataset is still limited, especially regarding grains of small size and of low trace element content. NanoSIMS measurements have started on different fractions of graphite grains [19].

The most studied group of presolar graphite grains is the KE3 fraction from the Murchison meteorite, which has a relatively low density of 1.65 – 1.72 g/cm^3, large size > 2 μm and typically high trace element content. Like for SiC grains of type X (Sec. 4.8.3) and silicon nitride grains, the presence of the radioactive nuclei ^{44}Ti [214], ^{41}Ca [16] and ^{49}V [281] at the time of the grain formation represents conclusive proof for an origin of low-density graphite grains in massive stars exploding as supernovæ of type II (SNII). The other isotopic signatures are similar to those shown by SiC-X grains and are, at least qualitatively, in agreement with a SNII origin: namely excesses in ^{15}N, ^{28}Si, ^{26}Al and, in addition to SiC-X grains, also excesses in ^{18}O. As shown in Fig. 6.2, where data from Ref. [281] are plotted, the ^{12}C/^{13}C ratios of graphite KE3 grains cover a very large range, from $\simeq$ 3 to $\simeq$ 10,000. The nitrogen isotopic ratios are typically lower than solar, indicating excesses of ^{15}N. The silicon composition of the grains, from Ref. [281], is illustrated in Fig. 6.3, showing ^{28}Si excesses up to twice the solar value, but also grains with excesses in ^{29}Si and ^{30}Si. In Figs. 6.2 and 6.3 the approximate compositions of the different regions of a SNII (see Fig. 2.5) are also indicated (from Ref. [299]). As in the case of SiC-X grains, some mixing is needed among the different layers of the SNII to match all the isotopic signature of low-density graphite grains.

Detailed studies of the possible results of such mixing and comparison with graphite data can be found in Refs. [299, 300, 281]. One difficulty relates to the question, briefly discussed in Secs. 1.5.1 and 4.8.3, of whether the condition C/O > 1 in the gas is necessary for the formation of graphite grains in SN ejecta. This condition is satisfied only in the outer regions of the star (those marked as **C/O**, **He/C** and **He/N** in Figs. 6.2 and 6.3), which have experienced H burning and partial He burning and are enriched

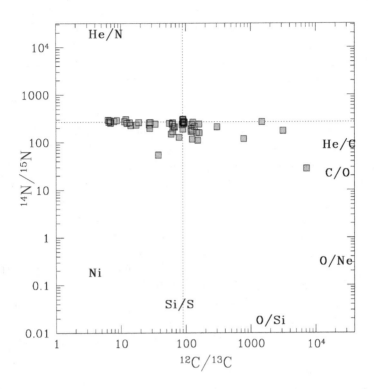

Fig. 6.2   Carbon and nitrogen isotopic compositions measured in single graphite grains from the KE3 fraction. Solar composition is indicated by the dotted lines. The approximate compositions of different regions of SNII are indicated by the labels, where, with reference to Fig. 2.5, **Ni** corresponds to the inner Fe- and Ni-rich region, **Si/S** to the region rich in ^{28}Si and ^{32}S, **O/Si** to the region rich in ^{16}O, ^{24}Mg and ^{28}Si, **O/Ne** to the region rich in ^{16}O, ^{20}Ne and ^{24}Mg, **C/O** to the region rich in ^{12}C and ^{16}O, **He/C** to the inner layer of the ^{4}He-rich zone, and **He/N** to the outer layer of the ^{4}He-rich zone. It is clear that the compositions of graphite grains are not matched by the composition of any single shell, and a mixture of them is needed to match the data.

in ^{18}O and ^{26}Al, as observed in the grains. On the other hand, the large amount of ^{44}Ti and ^{28}Si also observed in the grains, can be reproduced only by mixing of material from the inner regions of the star, where Si burning has taken place (Sec. 2.4). The problem is that, if C/O > 1 is needed, the mixing of outer and inner layers has to be accomplished limiting the contribution of the the intermediate oxygen-rich layers. Furthermore, as in the case of SiC-X grains, there are some problems in reproducing the observed abundance of ^{15}N and ^{29}Si.

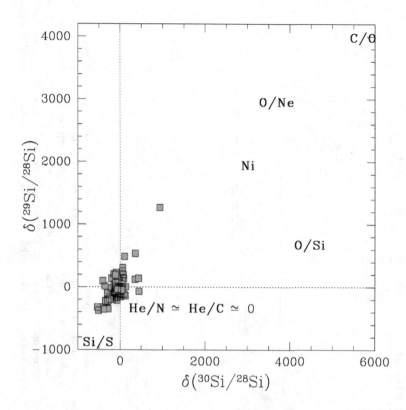

Fig. 6.3   Silicon isotopic compositions (in δ values) measured in single graphite grains from the KE3 fraction. Solar composition ($\delta = 0\ ^o/_{oo}$) is indicated by the dotted lines. The approximate compositions of different regions of SNII are indicated by the labels, in the same notation as in Fig. 6.2. Many grains have excesses in ^{28}Si, which can only be produced in material from the Si/S region.

6.3   Oxide grains

As introduced in Sec. 1.4, various types of presolar oxide grains have been recovered from primitive meteorites. These are corundum (Al_2O_3), spinel ($MgAl_2O_4$), hibonite ($CaAl_{12}O_{19}$) and titanium oxide (TiO_2) [58, 59, 136, 209, 211, 212, 213, 304, 307], as well as various types of silicate grains, both crystalline such as olivine ($Mg,Fe)_2SiO_4$ (where the bracket indicates a range of mixtures from pure Mg to pure Fe) and pyroxene $XY(Si,Al)_2O_6$ (where X and Y represents various other elements) and amorphous glass and GEMS (Glass with Embedded Metal and Sulfides)

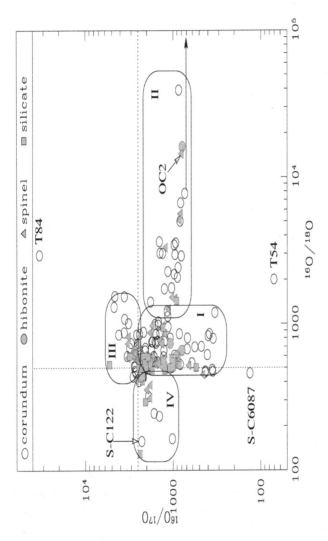

Fig. 6.4   Three-isotope plot showing the oxygen isotopic composition measured in oxide grains of the different types. Solar composition is indicated by the dotted lines. The four populations discussed in the text are enclosed in boxes and indicated by the labels I, II, III and IV. In one corundum grain of population II no ^{18}O could be detected, and only a lower limit from experimental uncertainties for the $^{16}O/^{18}O$ is available, which is indicated by the long horizontal arrow. The only titanium oxide grain discovered so far has ratios within 20% of solar and it is not included in the plot. Three corundum grains (T54, T84 and S-C6087) have compositions that cannot be classified within any of the four populations. Grain T84 almost certainly was produced in a SNII. Also for grain S-C122 a SNII origin has been proposed.

[188, 194, 195, 196, 199, 200]. The oxygen compositions of the grains are plotted in Fig. 6.4. The majority of such compositions cover the same region of the oxygen three-isotope plot, independently of the type of grain, indicating a common origin. Nittler *et al.* [212] presented a detailed study of the composition of oxide grains and in the following some of these authors' results are reported and explained in detail.

Oxygen has three stable isotopes with atomic mass number 16, 17 and 18. Of these, by far the most abundant is ^{16}O which is copiously produced in massive stars during both carbon and neon burning (see Sec. 2.3). Instead, ^{17}O and ^{18}O are much less abundant: in the solar system there is only 1 atom of ^{18}O per 500 atoms of ^{16}O, and 1 atom of ^{17}O per 2660 atoms of ^{16}O! The majority of the oxide grain data (those belonging to the populations I and II described below) are located in the oxygen three-isotope plot of Fig. 6.4 in the same region covered by spectroscopic observations of red giant and Asymptotic Giant Branch stars ([110, 111, 112, 152, 255], and see Fig. 9 of [212]), pointing to such an origin for the majority of oxide grains. This composition is in qualitative agreement with model predictions of the oxygen surface abundances of red giant and AGB stars, which are modified by the operation of the first dredge-up as discussed below. While carbonaceous phases can form only from gas where $C/O > 1$ (Sec. 1.4) and are observed to be present around carbon-rich evolved stars, oxide and silicate grains can form from gas where $C/O < 1$ and are observed to be present around oxygen-rich evolved stars [257]. Thus, oxide and SiC grain data together can produce a complete picture for our understanding of both the oxygen-rich and the carbon-rich phases of the evolution of red giant and AGB stars.

As is done for SiC grains (Sec. 4.1), a first step to interpret the composition of oxide grains is to classify them into different subgroups on the basis of their isotopic composition. Grains are separated into four different populations based on their position in the oxygen three-isotope plot. In Fig. 6.4 the areas covered by the different populations are enclosed in boxes. Note also that a fraction of corundum grains have excesses in ^{26}Mg. Since ^{26}Al is produced during the AGB phase (Sec. 4.5), but not during the earlier red giant phase, it is in principle possible to distinguish between corundum grains formed during the AGB phase, with ^{26}Mg excess, and grains that formed earlier, during the red giant phase, with no ^{26}Mg excess. If excesses of ^{26}Mg are present, the $^{26}Al/^{27}Al$ ratio can be derived in corundum grains. The four populations established on the basis of oxygen composition appear to have specific features also with regards to the aluminium composition.

Grains belonging to population I show $^{16}O/^{18}O$ ratios from solar to approximately a factor of two higher than solar and enhancements of ^{17}O. Their composition is compatible with an origin in red giant and AGB stars, where ^{17}O is enhanced at the stellar surface after the occurrence of the first dredge-up (Sec. 4.3). The highest observed ^{17}O enhancements, corresponding to the lowest $^{16}O/^{17}O$ ratios in Fig. 6.4, around 200, are attributable to stars of mass $\sim 3\ M_\odot$ (see Table 4.2). In fact, the effect of the first dredge-up increases with the stellar mass up to $\simeq 3\ M_\odot$, and then decreases again for stars of higher mass [41]. On the other hand, models of the first dredge-up do not predict the range of variation of a factor of $\simeq 2$ observed in the $^{16}O/^{18}O$ ratios of the grains of population I. Such variations are thus attributed to changes in the initial composition of the parent stars due to Galactic chemical evolution, in a similar fashion as suggested for the Si distribution in mainstream SiC grains (Sec. 4.6) [39, 41, 133, 212]. The ratio $^{26}Al/^{27}Al$ in corundum grains of population I has a mean value of 0.0023, in agreement with theoretical predictions for AGB stars [91, 92, 198, 290].

Oxide grains of population II show similar $^{16}O/^{17}O$ ratios as grains of population I. In addition they are depleted in ^{18}O: their $^{18}O/^{16}O$ ratios reach down to values two orders of magnitude lower than solar. In one corundum grain no ^{18}O could be detected so that only a lower limit for the $^{16}O/^{18}O$ is available. Such depletions can be explained by the operation of proton-capture processes at the base of the convective envelope in AGB stars. The grain compositions could be qualitatively associated with hot bottom burning occurring in intermediate-mass AGB stars (Sec. 4.2.1), however detailed calculations demonstrated that the composition of many grains cannot be quantitatively reproduced by hot bottom burning [42]. Such grains are better explained by extra-mixing processes in low mass stars during the AGB phase (*cool bottom processing*) [289] (Sec. 4.3). Nollett *et al.* [216] presented a detailed study where presolar grain data are compared to analytic models of cool bottom processing in AGB stars of low mass. They showed that the composition of the oxide grains of population II can be well explained by the occurrence of this type of extra mixing. The $^{26}Al/^{27}Al$ ratio in corundum grains of population II has a mean value approximately three times higher than that of grains of population I, which also indicates the operation of proton-capture processes. Oxide grains of population II represent the first direct evidence of the occurrence of extra mixing during the AGB phase, and can be used to resolve the large uncertainties related to this process.

The composition of grains belonging to populations III and IV, showing excesses and depletion of ^{16}O, respectively, can be interpreted as the result of Galactic chemical evolution of the oxygen isotopes. However, there are still large theoretical uncertainties. Oxygen-16 is produced in massive stars of all metallicities, starting from the ^{4}He initially present in the star, and thus its production is of primary nature, like that of ^{28}Si (see Sec. 4.6). On the other hand, ^{17}O and ^{18}O are produced starting from the initial CNO abundance, hence they depend on the initial metallicity of the star, and thus their production is of secondary nature, as that of ^{29}Si and ^{30}Si. Oxygen-17 is produced by the CNO cycle in all stars and carried to the surface by the occurrence of the first and second dredge-up. Moreover, this isotope is produced by the hot CNO cycle during nova outbursts. Oxygen-18 is produced in massive stars by α captures on ^{14}N from the H-burning ashes. Because ^{17}O and ^{18}O are secondary isotopes and ^{16}O is a primary isotope, the oxygen isotopic ratios ^{16}O/^{17}O and ^{16}O/^{18}O decrease with time in the Galaxy. This is qualitatively similar to the evolution of the Si isotopic ratios discussed in Sec. 4.6. Hence, grains belonging to populations III and IV could be explained as originating from stars of metallicities lower and higher than solar, respectively. However, this is quite an unsatisfactory explanation for grains belonging to population IV, as it carries the same paradox encountered for the Si isotopes in SiC grains, namely the fact that their parent stars should have been born before the Sun, but with a metallicity higher that solar (Sec. 4.6).

There is also the possibility that the ^{18}O excesses in grains of population IV are produced by AGB star nucleosynthesis, if ^{18}O is not completely converted into ^{22}Ne in the He intershell. Some grains of population IV have also been attributed to a SNII origin due their ^{18}O enhancements. In particular a possible SNII origin was discussed in detail for grain S-C122, even though the ^{16}O/^{17}O ratio and the Mg and Ti compositions of the grain do not support this hypothesis [58]. With regards to the ^{26}Al/^{27}Al ratios, those of grains belonging to population IV are similar to those of grains belonging to population I, while grains belonging to population III show ^{26}Al/^{27}Al ratios on average a factor of five smaller that those of populations I and IV.

The unique ^{16}O-rich oxide grain recovered to date, T84, almost certainly was produced in a SNII, since ^{16}O is the most abundant heavy isotope ejected by SNII [213]. It is still unclear why the abundance of oxide grains from SNII is so small since these types of grains are believed to be more likely to form in such environments than carbonaceous grains. Perhaps such grains are too small to be recovered from meteorites. The two corundum grains S-C6087 [59] and T54 [212] cannot be classified within any of the four populations and there is not yet a plausible explanation for their compositions.

As a final remark, the importance of the discovery and analysis of spinel and silicate grains for the understanding of the composition of red giant and AGB stars should be emphasised. For each of these grains it is in principle possible to derive the compositions of both oxygen and magnesium (in the case of spinel [304, 307]), or both oxygen and silicon (in the case of silicate grains [194, 199]). For example, one spinel grain (OC2) was found to have an oxygen composition typical of population II, but excesses in both ^{25}Mg and ^{26}Mg. This is the first time the composition of a presolar grain indicates an origin from oxygen-rich intermediate-mass AGB stars, and this conclusion could only been derived because its Mg isotopic composition could be measured. This grain sets constraints both on the modelling of convection in intermediate-mass AGB stars, which controls the effect of third dredge-up and the operation of hot bottom burning, and on the nuclear reaction rates that determine the abundances of the O and Mg isotopes.

Moreover, the composition of oxygen, magnesium and silicon are all believed to have been somewhat affected by the chemical evolution of the Galaxy. Hence, the possibility to consider together, for example, oxygen and silicon isotopic ratios allows us to test interpretative hypotheses for grain compositions proposed to date and to set strong constraints on our understanding of the Galactic chemical evolution of the isotopes of such elements. For example, four pyroxene grains are reported to span a range from 129 to 943 in their ^{16}O/^{18}O ratios [194]. However, their silicon isotopic ratios are all within $\simeq 5\%$ of solar composition. This makes it difficult to attribute the ^{16}O/^{18}O variations to the effect of Galactic chemical evolution, which should also change the silicon composition. The dataset is still limited, but its extension in the near future will produce rapid advancements: no other observations allow theories of the origin of the elements to be tested to the same level of precision.

6.4   Exercises

(1) Show that, if within approximately two hours Xe is separated from I, the isotopic pattern of Xe-H can be reproduced.

(2) Theoretical models predict that variation in the ^{17}O/^{16}O and ^{18}O/^{16}O

ratios in the Galaxy with time are roughly proportional to variations of the metallicity [279]. What is the range of metallicity of the parent stars of oxide grains of populations III and IV, if we assume that their oxygen isotopic compositions are due to the effect of the Galactic chemical evolution? Is this solution realistic? How does it compare with the results obtained for Exercise 4.9.3?

Appendix A

Glossary

α **particle** An ^{4}He nucleus.

α **process** Used to indicate C, Ne and O burnings, by which nuclei with atomic mass number $A = integer \times 4$ (α nuclei) up to ^{32}S are produced.

β **particle** An electron $(-)$ or a positron $(+)$ emitted during a **radioactive** decay.

δ **notation** $\delta(x/y)$ represents the deviation of the x/y ratio from the solar system ratio $(x/y)_\odot$: $\frac{(x/y)-(x/y)_\odot}{(x/y)_\odot}$, multiplied by one thousand, i.e. permil $(^o/_{oo})$.

γ **ray** High-energy photon.

ν **process Nucleosynthetic** process occurring during a **supernova** of type II, where neutrinos coming from the nascent neutron star bombard nuclei, resulting in the ejection of **nucleons** by **spallation** reactions.

Alpha-rich freeze out α-capture reactions occurring in the O- and Si-rich regions of a **supernova** of type II, when the temperature drops after the shock wave and the gas moves out of equilibrium (see the e **process**), resulting in a significant final abundance of α nuclei.

Anomalous Usually refers to an **isotopic** composition differing from that displayed by the bulk of the material in the solar system; also referred to as exotic.

Asymptotic Giant Branch (AGB) An advanced phase, lasting approximately a million years, of the life of a star of initial mass lower than approximately 10 $M_\odot$, during which thermal runaways periodically occur in the He-rich region located at the top of the **degenerate** C-O core. Carbon is periodically carried to the surface by third **dredge-up** and the star can become richer in carbon than in oxygen. The H-rich stellar envelope is progressively reduced by strong stellar winds. During the post-AGB phase the last fraction of the envelope is lost and a planetary nebula is formed, with a **white dwarf** in the middle.

Bulk When used in the expression "measurements of grains in bulk" and similar, indicates that the laboratory analysis of the grains has been carried on a collection of a large number of grains. Measurements in bulk yield very precise *average* information (as opposed to measurements in single grains, which yield specific information on each grain), but carry the risk that the sample may be polluted by contaminants.

Convection The transportation of energy in stars by the movement of the stellar gas, as opposed to **radiative transport**.

Cool bottom processing Nucleosynthesis and mixing process that can occur in **red giant** and **Asymptotic Giant Branch** stars, by which material of the **convective** H-rich envelope suffers proton-capture reactions when it is carried to high-temperature regions below the base of the envelope.

Cosmic rays High-energy particles, typically ions (mostly protons but also α **particles** and heavier nuclei), that travel across the Galaxy and probably have been accelerated by **supernova** shocks (Galactic cosmic rays), or originating from the Sun and travelling across the solar system (solar cosmic rays).

Degenerate gas High-density electron (in **white dwarves**) or neutron (in **neutron stars**) gas where pressure increases not because the temperature increases but because particles are forced to occupy quantum-mechanical states of higher and higher energy because of the Pauli exclusion principle.

Deuterium The very rare heavy isotope of hydrogen ^{2}H, or D, made up of one proton and one neutron. Only 0.015% of hydrogen in the solar system is made up of deuterium.

Dredge-up Mixing mechanism occurring episodically in **red giant** and **Asymptotic Giant Branch** stars, by which material processed by **nuclear reactions** is carried from the deep layers of the star to the surface. The *first* dredge-up occurs when H is exhausted in the core and the star moves to the red giant branch, the *second* dredge-up occurs when He is exhausted in the core and the star moves to the Asymptotic Giant Branch. *Third* dredge-up is the collective name for the recurrent dredge-up episodes during the Asymptotic Giant Branch phase.

e **process** Nuclear statistical *equilibrium* process during which stellar material behaves as a thermodynamic gas trying to reach and maintain *equilibrium* by means of **nuclear reactions**. In this process the formation of the most stable nuclei is favoured, so that the main result is to convert all the initial material into nuclei belonging to the Fe peak.

Extra mixing Generally used to refer to processes in stars when mixing occurs beyond the boundary of the **convective** regions defined by the Schwarzschild criterion.

Fractionation Separation of different elements, or different isotopes, in a material due to a preference in the process of their incorporation.

Galactic Chemical Evolution (GCE) The changes of the chemical composition within a galaxy at different times and locations due to the continuous cycle by which material in the galaxy is processed by nucleosynthesis inside stars and then ejected into the **interstellar medium** from which new stars are born.

Helium burning The triple-α reaction by which three α particles are fused into ^{12}C, and the subsequent α capture on ^{12}C that leads to the production of ^{16}O.

Hot bottom burning Nucleosynthesis and mixing process that can occur in **Asymptotic Giant Branch** stars, by which material of the **convective** H-rich envelope suffers proton-capture reactions when it is carried to high-temperature regions at the base of the envelope.

Hydrogen burning Generally indicates various sequences of reactions occurring in the presence of protons, such as the pp chain and the CNO, NeNa and MgAl cycles.

Implantation The process by which ions of volatile elements can be inserted and embedded in dust grains.

Interstellar Medium (ISM) Material permeating space in between stars, composed of 99% gas and 1% cosmic dust.

Interplanetary Dust Particles (IDPs) Tiny meteorites (diameter < 50 μm) collected in the upper region of the atmosphere of the Earth.

Isotopes Nuclei with the same number of protons (Z), but differing numbers of neutrons (N). Isotopes of a given element share the same chemical properties, but are identified by different mass numbers $(A = Z + N)$. For example, carbon has two stable isotopes indicated as ^{12}C and ^{13}C, which have the same number of protons $(Z = 6)$, but different numbers of neutrons: ^{12}C has $N = 6$ and ^{13}C has $N = 7$, so that their atomic mass numbers are 12 and 13 respectively. The isotopic composition of an element is a product of nucleosynthesis and therefore a very important property to be measured in meteoritic grains. Isotopic ratios, such as ^{12}C/^{13}C, measured in presolar grains routinely deviate by orders of magnitude from the ratios observed in the bulk of solar-system material, and represent the indication of their stellar origin.

Magic numbers The numbers 2, 8, 20, 28, 50, 82, for which energy levels for protons and/or neutrons are filled in a nucleus, thus making the structure extremely stable against **nuclear reactions**.

Mass spectrometer Instrument designed to extract ions from a sample, separate them depending on their mass and measure the number of ions of each mass. Widely used in the laboratory analysis of stellar dust grains.

Metallicity The fraction of mass composed by elements heavier than He, also commonly indicated by Z. For example, about 2% of the material present in the Sun is composed of elements heavier than He, thus $Z_\odot \simeq 0.02$.

Meteorite A fragment of rock that has reached the surface of the Earth from space. Meteorites are classified as stony meteorites, iron meteorites, and stony-iron meteorites. Among stony meteorites, Chondrites are characterised by *chondrules*: small spheres (average diameter of 1 mm) of formerly melted minerals that have come together with other mineral matter to form a solid rock. Chondrites are believed to be among the oldest rocks in the solar system and contain presolar stellar grains. *Carbonaceous* chondrites, such as the Murchison meteorite, contain carbon also in the form of organic compounds.

Main sequence The first and longest phase of the life of a star (where the Sun is now) during which **H burning** in the centre produces the nuclear energy that sustains the structure against gravitational collapse.

Molecular cloud Cool and dense region of the **interstellar medium** where atoms tend to be combined into molecules, and where the formation of stars is more likely to occur.

Neutron star A very compact and dense object with a gas of neutrons in **degenerate** conditions.

Nova A star that suddenly increases in brightness and then decreases to its former brightness due to thermonuclear runaways, with associated **H burning**, periodically occurring when hydrogen is accreted from a companion onto the surface of a **white dwarf**.

Noble gas Noble gases are He, Ne, Ar, Kr and Xe. They have a very stable atomic structure and hence do not react easily with other atoms. They are also the most volatile elements, condensing only at extremely low temperatures, and are known as *rare* gases as their concentrations in other materials are very low. However, they can be found as trapped or implanted bubbles inside materials. Analysis of their isotopic composition in primitive meteorites led to the discovery of presolar stellar grains.

Nucleon A particle composing a nucleus, either a proton or a neutron.

Nuclide Chart A convenient way to represent nuclei, as a Periodic Table is for the elements. In a nuclide chart, or Chart of the Nuclides, each isotope is located as a function of its number of protons (Z, on the y-axis) and number of neutrons (N, on the x-axis). Thus, increasing the value on the y-axis heavier and heavier elements are drawn, while increasing the value on the x-axis the isotopes of each element become richer in neutrons. Isotopes with the same atomic mass number $A = Z + N$ result to be plotted along diagonal lines. Stable and unstable isotopes are usually distinguished by different means of representations, and pertinent information is also included.

Nucleosynthesis The formation of nuclei via **nuclear reactions**. Nuclear reactions are activated inside stars where temperatures reach up to billions of degrees (stellar nucleosynthesis). Only hydrogen and a fraction of helium were produced in the early phases of the Universe, thus the first stars contained only H and He. Nuclei that were produced by the first generation of stars, i.e. through nucleosynthesis starting only from H and He, are called *primary* nuclei. Nuclei that are produced only if elements heavier than He are also present in the star are called *secondary* nuclei.

Nuclear reactions Fusion or fission of nuclei, governed by electromagnetic and nuclear interactions.

Overshoot The extension of motion from **convective** regions into **radiative** regions where the material, unable to completely brake at the Schwarzschild border, changes its state from motion to stability.

Photodisintegration Reaction by which a nucleus is fragmented by interaction with a high energy photon (γ **ray**).

p **process** *Proton*-capture process producing a very small fraction of the Galactic abundances of elements heavier than iron, in particular those of isotopes that cannot be produced by neutron-capture processes (the *r* and the *s* processes), which are also known as *p*-only isotopes.

Presolar Material that survived the homogenisation of the bulk of material during the formation of the solar system, and thus has preserved an **anomalous** isotopic signature that predates the formation of the Sun.

Radioactive A nucleus which undergoes spontaneous nuclear decay involving emission of radiation in the form of α, β, γ **particles** and neutrinos. Some nuclei have long half-lives against decay and can be referred to as long-living (e.g. ^{26}Al, with half-life of $\simeq 0.7$ million years) or very long-living (e.g. ^{238}U, with half-life of $\simeq 4.5$ billion years) isotopes.

Radiative transport The transportation of energy in stars occurring by means of electromagnetic radiation, as opposed to **convection**.

Red giant Stellar evolutionary phase following the **main sequence** phase, where a star expands by more than 100 times its initial size and cools at the surface, as a consequence of the exhaustion of H in the centre, its contraction, and the onset of H burning in a shell around the centre.

Refractory Elements that condense from a gas into a solid at high temperatures, i.e. early in the condensation sequence in a cooling gas. Examples of refractory elements are Zr, Mo and Nd. They have condensed into stellar grains during their formation at high temperatures.

r process *Rapid* neutron-capture process (as opposed to the *s* process) occurring at relatively high neutron densities, $\simeq 10^{20}$ cm^{-3}, and responsible for the production of about half of the Galactic abundances of elements heavier than iron. Isotopes that can only by produced by the *r* process are named *r* only.

Semiconvection When, because of mixing due to convection, the composition of the material in a region changes, producing a change in the opacity and making the region radiatively stable, thus stopping the mixing.

Spallation reaction When a nucleus is hit by a very high energy particle and smashed into many fragments.

Spectroscopy Applied to stars, the investigation of the spectra, i.e. the pattern of absorption or emission of electromagnetic radiation from stars. Spectra analysis reveals the chemical composition of a stellar surface, which is also a function of the temperature of the star, and has been used to classify stars into different spectral classes.

s process *Slow* neutron-capture process (as opposed to *r* **process**) occurring at relatively low neutron densities, $\simeq 10^8$ cm^{-3}, and responsible for the production of about half the Galactic abundances of elements heavier than iron. Isotopes that can only by produced by the *s* process are named *s* only.

Supernova A star that undergoes a sudden and temporary increase in brightness to a much greater degree than a **nova**, as a result of an explosion. Supernovæ that show the presence of H in their spectra are classified as type II (SNII) and are the product of exploding massive stars (of mass higher than approximately 10 $M_\odot$). Supernovæ that do not show the presence of H in their spectra are classified as type I (SNI) and are the result of a **white dwarf** accreting material from a companion (SNIa) or of the explosion of **Wolf-Rayet stars** (SNIb,c).

Three-isotope plot Graph in which two isotopic ratios, with a common reference isotope, are plotted against one another. Three-isotope plots are widely used when comparing isotopic ratios because a composition resulting from the mixing of two different components lies on a straight line connecting the two components and the degree of mixing between the two components can be estimated by the position of the point on the mixing line.

Volatile Elements that condense from gas into solid at low temperature, i.e. late in the condensation sequence in a cooling gas. The most volatile elements are the **noble gases**, which condense only at very low temperature. They have been **implanted** into stellar grains after being ionised.

White dwarf A very compact and dense object with a gas of electrons in **degenerate** conditions.

Wolf-Rayet stars Stars initially very massive (masses higher than approximately 50 $M_\odot$) that lose all their H-rich envelope because of strong stellar winds before undergoing a **supernova** explosion.

x process The x-process refers to Big Bang nucleosynthesis as well as interstellar nucleosynthesis due to **spallation** reactions, and is believed to be responsible for the production of the light elements of low abundance: ^{2}H, Li, Be, B.

Appendix B

Solutions to Exercises

B.1   Chapter 1

(1) The radius of an atom is about 1 Angstrom $= 1$ Å $= 10^{-10}$ m and its volume is proportional to 10^{-30} m^3. The volume of a presolar grain of radius 0.01 μm $= 10^{-8}$ m is proportional to 10^{-24} m^3. Hence, approximately $10^{-24}/10^{-30} = 10^6$ atoms are present. If the radius of the grain is 1 μm, then approximately 10^{12} atoms are present.

(2) One gram of the Murchison meteorite corresponds to 1000 milligrams. Since diamonds are present with an abundance of ~ 750 part per million in mass (Table 1.2) there are $\sim 7.5 \times 10^{-4}$ milligrams of diamond in 1 milligram of Murchison. i.e. 0.75 milligrams of diamond in 1 gram of Murchison. Equivalently, there are $\sim 9 \times 10^{-3}$ milligrams of SiC and $\sim 10^{-3}$ milligrams of graphite.

(3) The typical size of a diamond grain is 2 nm $= 2 \times 10^{-9}$ m (Table 1.2). Hence, there are approximately 10,000 atoms in each diamond (see Exercise 1) and ten billion atoms would constitute about 10^6 diamond grains. Since about one nanodiamond in each million contains an atom of anomalous Xe (Sec. 1.4), only approximately one atom of anomalous Xe is present in ten billion atoms.

B.2   Chapter 2

(1) Let us consider the fraction of mass of the species i, $X_i = M_i/M$. There $M = \rho V$ and $M_i =$ the number of particle of species i multiplied by the mass of each of them $= N_i A_i m_u$, where A_i is the atomic mass of

species i and m_u atomic mass unit. Hence:

$$X_i = \frac{M_i}{M} = \frac{N_i A_i m_u}{\rho V} = \frac{n_i A_i m_u}{\rho},$$

since $n_i = N_i/V$. But $n_i m_u/\rho = Y_i$, thus $X_i = A_i Y_i$.

(2) The dimensions of the rate $n_i n_j \langle \sigma v \rangle$ are $[L^{-3} \ L^{-3} \ L^2 \ L^1 \ T^{-1}]$, where L is a unit of length and T is a unit of time. Simplifying one obtains $[L^{-3} \ T^{-1}]$, as required.

Given that $n_i = \rho Y_i/m_u$, in terms of Y one gets:

$$n_i n_j \langle \sigma v \rangle = \frac{Y_i Y_j \rho^2}{m_u^2} \langle \sigma v \rangle.$$

In terms of X:

$$n_i n_j \langle \sigma v \rangle = \frac{X_i X_j \rho^2}{A_i A_j m_u^2} \langle \sigma v \rangle.$$

(3)

$$\eta = \sum_i (N_i - Z_i) Y_i = \sum_i (N_i - Z_i)\frac{X_i}{A_i} =$$

$$\sum_i (N_i - Z_i + Z_i - Z_i)\frac{X_i}{A_i} = \sum_i (A_i - 2Z_i)\frac{X_i}{A_i} =$$

$$\sum_i A_i \frac{X_i}{A_i} - \sum_i 2Z_i \frac{X_i}{A_i} =$$

$$1 - 2\sum_i \frac{Z_i}{A_i} X_i = 1 - 2Y_e.$$

Assume that all the metallicity is represented by CNO nuclei, which are all converted into ^{22}Ne so that after H and He burning $X_{22} = 0.02$, $N(^{22}\text{Ne}) = 12$ and $Z(^{22}\text{Ne}) = 10$ and that all the other nuclei present in the material have $N_i = Z_i$, i.e. $Z_i/A_i = 0.5$.

$$\eta = 1 - 2Y_e = 1 - 2\sum_i \frac{Z_i}{A_i} X_i = 1 - 2 \times (0.5 \times 0.98 + 10/22 \times 0.02) = 0.0018,$$

alternatively:

$$\eta = \sum_i (N_i - Z_i)Y_i = \frac{12-10}{22}0.02 = 0.0018.$$

(4) See picture in the next page.
(5) The branching factor is

$$f_n = \frac{\lambda_n}{\lambda_n + \lambda_\beta},$$

where, in the case of ^{95}Zr (1 mbarn $= 10^{-27}$ cm^2):

$$\lambda_n = N_n \langle \sigma \rangle v_{th} = N_n \times 60 \times 10^{-27} \times 2.1 \times 10^8 = N_n \times 1.26 \times 10^{-17},$$

and

$$\lambda_\beta = ln\, 2/T_{1/2} = \frac{0.693}{64 \times 8.64 \times 10^4} = 1.25 \times 10^{-7}.$$

Thus,

$$f_n = \frac{N_n \times 1.26 \times 10^{-17}}{N_n \times 1.26 \times 10^{-17} + 1.25 \times 10^{-7}},$$

which for $N_n = 5 \times 10^7, 5 \times 10^8, 5 \times 10^9$ and 5×10^{10} cm^{-3} gives 0.005, 0.05, 0.33, 0.83, respectively.

B.3 Chapter 3

(1) The mass of a nucleus is calculated as $A \times m_u + \Delta$, or equivalently, $A \times E_u + \Delta$, where Δ is the mass excess, A is the atomic mass number and m_u is the atomic mass unit $= 1.66 \times 10^{-27}$ Kg, so that $E_u = m_u c^2$ $= 1.49 \times 10^{-10}$ J $= 931$ MeV. Thus, the mass of ^{17}O: $= 17 \times 931 -$ $0.809 = 15,826$ MeV, i.e. $16.999\, m_u$. The mass of ^{16}O is $15.995\, m_u$ and that of H is $1.008\, m_u$ (corresponding to the mass of a proton plus an electron). (Note that the atomic weight of the elements reported in a periodic table of the elements are different from those calculated here as they are weighted by the solar abundance of each isotope). So the mass of a ^{16}OH ion is $\simeq 17.003\, m_u$. The mass resolution needed to separate the ions is then $\simeq 5000$.

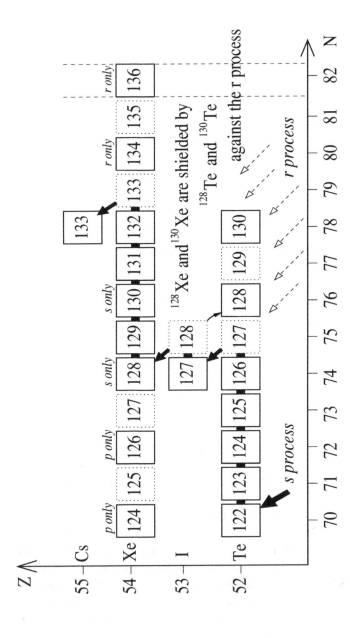

Fig. B.1 Section of the nuclide chart including some isotopes of elements from Te to Ce. The nucleosynthetic signature of Xe-S and Xe-HL illustrated in Figs. 1.3 are easily interpreted as *s*-process and *r, p*-process signatures, respectively. In the case of Xe-S the *s*-only isotopes ^{128}Xe and ^{130}Xe are the most abundant isotopes (with respect to solar composition). In the case of Xe-HL the *p*-only and *r*-only isotopes, ^{124,126}Xe and ^{134,136}Xe, respectively, are instead enhanced.

(2) In an electric field ions have a kinetic energy $E_{kin} = qV = 1/2\, m\, v^2 = 1/2\, m\, d^2/t^2$, where d is the distance and t the time (Sec. 3.3). Thus:

$$\frac{t_{17O}}{t_{16OH}} = \sqrt{\frac{m_{17O}}{m_{16OH}}}$$

and the difference in transit time is of 0.012%.

In a magnetic field ions have an orbit of radius $r = \frac{mv}{qB}$ (Sec. 3.3), where the velocity v is defined by the previously applied voltage V (see above). Thus:

$$\frac{r_{17O}}{r_{16OH}} = \sqrt{\frac{m_{17O}}{m_{16OH}}}.$$

(3) Since the energy of the particles is all in the form of kinetic energy then $1/2\, m\, v^2 = 20$ KeV, where m is the mass of the particles and v is their speed. Hence, given that $m = 133\, m_u = 133 \times 1.66 \times 10^{-27}$ Kg and 20 KeV $= 3.20 \times 10^{-15}$ J, the speed $\simeq 1.7 \times 10^5$ m/s.

B.4   Chapter 4

(1)   a)

$$1 < \frac{^{12}C}{^{16}O} = \frac{16}{12}\frac{X(^{12}C)}{X(^{16}O)} =$$

$$\frac{16}{12}\frac{X(^{12}C)_{initial}\,M_{env} + X(^{12}C)_{intershell}\,M_{TDU}}{X(^{16}O)_{initial}\,M_{env} + X(^{16}O)_{intershell}\,M_{TDU}} =$$

$$\frac{16}{12}\left(\frac{0.002\,M_{env} + 0.23\,M_{TDU}}{0.009\,M_{env}}\right) = \frac{16}{12}\left(\frac{0.002}{0.009} + \frac{0.23}{0.009}\,DIL\right).$$

$$DIL > 0.020,\ 1/DIL < 50.$$

b) For a metallicity 1/3 of solar DIL = 0.0069, 1/DIL = 145. It is three times easier in this case to produce a carbon-rich star than in the solar case because there is three times less initial oxygen abundance to overtake.

c) Applying the same formula as above:

$$\frac{^{12}\mathrm{C}}{^{13}\mathrm{C}} = \frac{13}{12} \frac{X(^{12}\mathrm{C})_{initial}\, M_{env} + X(^{12}\mathrm{C})_{intershell}\, M_{TDU}}{X(^{13}\mathrm{C})_{initial}\, M_{env} + X(^{13}\mathrm{O})_{intershell}\, M_{TDU}}.$$

For solar metallicity:

$$\frac{13}{12}\left(\frac{0.002\, M_{env} + 0.23\, M_{TDU}}{0.0002\, M_{env}}\right) = \frac{13}{12}\left(\frac{0.002}{0.0002} + \frac{0.23}{0.0002}\, DIL\right) = 36.$$

For metallicity one third of solar:

$$\frac{13}{12}\left(\frac{0.00067}{0.000067} + \frac{0.23}{0.000067}\, DIL\right) = 36.$$

(2)

$$\frac{^{20}\mathrm{Ne}}{^{22}\mathrm{Ne}} = \frac{22}{20}\frac{X(^{20}\mathrm{Ne})}{X(^{22}\mathrm{Ne})} = \frac{22}{20}\frac{X(^{20}\mathrm{Ne})_\odot \times Z/Z_\odot}{Z} = 0.089$$

for any value of Z.

$$\frac{^{4}\mathrm{He}}{^{22}\mathrm{Ne}} = \frac{22}{4}\frac{X(^{4}\mathrm{He})}{X(^{22}\mathrm{Ne})} = \frac{22}{4}\frac{0.7}{Z},$$

thus $^{4}\mathrm{He}/^{22}\mathrm{Ne}=192.5$ for $Z = Z_\odot = 0.02$, 385 for $Z = Z_\odot/2 = 0.01$ and 641 for $Z = Z_\odot/3 \simeq 0.006$.

(3)   a) Mainstream SiC grains have $^{29}\mathrm{Si}/^{28}\mathrm{Si}$ ranging from 0.94 to 1.19 of the solar ratios (Fig. 4.9). Thus the range in metallicity will be from 0.88 to 1.42 $Z_\odot$, i.e. from 0.018 to 0.028.

   b) A simple average for the metallicity of the mainstream SiC grain parent stars is 1.15 $Z_\odot$, i.e. 0.023. For Y grains the $^{29}\mathrm{Si}/^{28}\mathrm{Si}$ range is from $\simeq 0.95$ to $\simeq 1.10$ (Fig. 4.2), corresponding to an average metallicity of 1.025 $Z_\odot$, i.e. 0.021. For Z grains the $^{29}\mathrm{Si}/^{28}\mathrm{Si}$ range is from $\simeq 0.84$ to $\simeq 1.07$, corresponding to an average metallicity of 0.955 $Z_\odot$, i.e. 0.019. Thus, $Z_{mainstream} > Z_Y > Z_Z$. This result is in qualitative agreement with the discussion presented in Sec. 4.8.1, i.e. with the fact that Y and Z grains are believed to have originated in AGB stars of metallicity lower than solar. However, stellar models of metallicities $Z \sim 0.5 Z_\odot$ and $Z \sim 0.3 Z_\odot$, i.e. much lower than those obtained by the GCE models, are needed to produce the ^{30}Si/^{28}Si ratios observed in Y and Z grains, respectively. The problem is open to further study.

c) Following from their metallicity range, the parent stars of main-stream SiC grains should have been born between 9 and 15 billion years, i.e. from 1 billion years before the Sun, to 5 billion years after the Sun. This is not realistic because SiC grains should have been produced before the solar birth in order to be included in the protosolar nebula.

B.5 Chapter 5

(1) a) At 100 million degrees the ^{13}C$(\alpha, n)^{16}$O and ^{22}Ne$(\alpha, n)^{25}$Mg reaction rates $N_A \langle \sigma v \rangle$ are $\simeq 10^{-13}$ and 3.2×10^{-34} cm^3 s^{-1} mole $^{-1}$, respectively. Since $N_A = 6.022 \times 10^{23}$ mole $^{-1}$, the Maxwellian-averaged cross section times relative velocity $\langle \sigma v \rangle$ are 1.66×10^{-37} and 5.3×10^{-58} cm^3 s^{-1}, respectively. For ^4He in He intershell conditions (see Exercise 2.6.1):

$$n(^4\text{He}) = \frac{X(^4\text{He})\,\rho}{4\,m_u} = \frac{0.7 \times 10^4}{4 \times 1.66 \times 10^{-24}} = 1.05 \times 10^{27} \text{cm}^{-3}$$

Thus, $\tau(^{13}$C$) = 5.7 \times 10^9$ s $= 181$ yr and the time it takes to reduce the ^{13}C abundance by a factor of 10 is 416 yr. On the other hand, $\tau(^{22}$Ne$) = 1.8 \times 10^{30}$ s $= 5.7 \times 10^{22}$ yr and the time it takes to reduce the ^{22}Ne abundance by a factor of 10 is 1.3×10^{23} yr. The ^{22}Ne$(\alpha, n)^{25}$Mg reaction is obviously not at work in such conditions.

At 300 million degrees the ^{13}C$(\alpha, n)^{16}$O and ^{22}Ne$(\alpha, n)^{25}$Mg reaction rates $N_A \langle \sigma v \rangle$ are $\simeq 3.2 \times 10^{-5}$ and 3.2×10^{-11} cm^3 s^{-1} mole $^{-1}$, respectively. Thus the time it takes to reduce the ^{13}C and the ^{22}Ne abundances by a factor of 10 is 7 minutes and 13 yr, respectively.

b) At 100 million degrees during the interpulse period the ^{13}C$(\alpha, n)^{16}$O reaction has enough time to be activated and completely destroy any ^{13}C. At 300 million degrees during the thermal pulse the ^{22}Ne$(\alpha, n)^{25}$Mg reaction is only marginally activated, as time is only enough to destroy a few percent of ^{22}Ne.

(2) Using the formula from Sec. 5.2.1 with $\delta = -700$:

$$\frac{\text{Mo}_G}{\text{Mo}_N} = \frac{\sqrt{2|\delta|^2}}{\sqrt{2|\delta + 1000|^2}} = 2.33$$

Since $\text{Mo}_{total} = \text{Mo}_G + \text{Mo}_N$, $\simeq 30\%$ of total Mo is from the N component, and the remaining 70% from the G component.

(3) For the two Xe isotopes in local equilibrium $^{128}\text{Xe}/^{130}\text{Xe} \simeq \langle\sigma_{130}\rangle/\langle\sigma_{128}\rangle$. With the neutron-capture cross sections recommended by Bao et al. [26] the ratio is 0.57, 30% higher than the value observed in SiC grains. With the neutron-capture cross sections of Reifarth et al. [231] the ratio is 0.50, 14% higher than the value observed in SiC grains. In Fig. 5.12 the predictions appear to be only 2% higher than the value observed in SiC grains. This is due to the presence of a small but interesting branching point at ^{128}I (see Fig. B.1), for which isotope the two β decay channels, + and − compete leading to a small bypass of ^{128}Xe with respect to ^{130}Xe [232].

(4) For the case with $\tau = 0.2$ mbarn^{-1}, $\langle\sigma_{88}\rangle N_{88}/\langle\sigma_{86}\rangle N_{86} \simeq 0.2$. Thus $N_{88}/N_{86} = 0.2 \times \langle\sigma_{86}\rangle/\langle\sigma_{88}\rangle = 2.$, $\frac{N_{88}/N_{86}}{(N_{88}/N_{86})_\odot} = 0.24$ and thus $\delta = -760$. When $\tau = 0.4$ mbarn^{-1}, $\langle\sigma_{88}\rangle N_{88}/\langle\sigma_{86}\rangle N_{86} \simeq 1.2$ and $\delta = 438$. When $\tau = 0.9$ mbarn^{-1}, $\langle\sigma_{88}\rangle N_{88}/\langle\sigma_{86}\rangle N_{86} \simeq 4.$ and $\delta = 3,800$. When $\tau = 3.8$ mbarn^{-1}, $\langle\sigma_{88}\rangle N_{88}/\langle\sigma_{86}\rangle N_{86} \simeq 1.3$ and $\delta = 559$. Only the result for $\tau = 0.4$ mbarn^{-1} would plot within the range of the y-axis of Fig. 5.7. The theoretical results shown in the plot are obtained for a neutron exposure with average values between $\simeq 0.1$ and $\simeq 0.4$ mbarn^{-1} from the ^{13}C pocket. However, one should keep in mind that detailed model predictions are complicated by the presence of the branching points at ^{85}Kr and ^{86}Rb (see Fig. 5.14).

(5) The branching factor is as from Exercise 2.6.5. In the case of ^{85}Kr and for the given neutron density:

$$\lambda_n = N_n\langle\sigma\rangle v_{th} = 4 \times 10^7 \times 1.4 \times 10^{-17} = 5.7 \times 10^{-10}\text{s}^{-1}.$$

and

$$\lambda_\beta = ln\,2/T_{1/2} = 2. \times 10^{-9}\text{s}^{-1}.$$

Thus, $f_n = 0.22$.

B.6   Chapter 6

(1) The isotopic pattern of Xe-H shows that $(^{134}Xe/^{136}Xe)_H = 0.75$ $(^{134}Xe/^{136}Xe)_\odot$ (Fig. 1.3). Assuming that the solar abundances are produced by pure r process, and that ^{136}Xe in Xe-H is also produced by the r process, then we obtain that $^{134}Xe_H/^{134}Xe_{r\,process}=0.75$. This means that the abundance of ^{134}I produced by the r process had only enough time to be reduced of a factor of 0.25 into ^{134}Xe. The half-life of ^{134}I is 52 minutes, $\tau = $ half-life $/ln2$, hence, $0.25=e^{-time/75m}$, from which $time \simeq 100m$.

(2) The most extreme oxide grain belonging to population III has $^{16}O/^{17}O$ and $^{16}O/^{18}O$ ratios roughly double the solar values, thus the range of metallicity covered by the parent stars of population III would be down to about half solar. The most extreme oxide grain belonging to population IV has $^{16}O/^{17}O$ and $^{16}O/^{18}O$ ratios roughly half the solar values, thus the range of metallicity covered by the parent stars of population III would be up to about double the solar value.

This range is larger than that obtained for the parent stars of SiC grains from their Si isotopic composition. Using the relation given in Exercise 4.9.3 between metallicity and time of birth of a star, one find that the parent stars of oxide grains should have been born between 5 and 22 billion years, i.e. from 5 billion years before the Sun (for population III grains), to 12 billion years after the Sun (for population IV grains). This conclusion for grains of population IV is even more puzzling than that reached for the parent stars of presolar SiC grains, using this simple approach.

Appendix C

Selected Books and Reviews for Quick Reference

C.1 Presolar grains

(1) Anders, E., and Zinner, E. (1993). Interstellar grains in primitive meteorites — Diamond, silicon carbide, and graphite *Meteoritics* **28**, pp. 490–514

(2) Bernatowicz, T. J., and Zinner, E., eds. (1997). Astrophysical Implications of the Laboratory Study of Presolar Materials *American Institute of Physics Conference Proceedings* **402**

(3) Zinner, E. (1998). Stellar Nucleosynthesis and the Isotopic Composition of Presolar Grains from Primitive Meteorites *Annual Review of Earth and Planetary Sciences* **26**, pp. 147–188

(4) Zinner, E. (2004). Presolar grains, in Treatise on Geochemistry. 1. Edited by K. K., Turekian, H. D. Holland, and A. D. Davis *Elsevier*, Oxford and San Diego, pp. 17–39

(5) Clayton, D. D., and Nittler, L. R. (2004). Astrophysics with Presolar Stardust *Annual Review of Astronomy and Astrophysics* **42**, pp. 39–78

(6) Lodders, K., and Amari, S. (2005). Presolar grains from meteorites: Remnants from the early times of the solar system *Chemie der Erde*, to appear

C.2 Stellar evolution and nucleosynthesis

(1) Burbidge, E. M., Burbidge, G. R., Fowler, W. A., and Hoyle, F. (1957). Synthesis of the Elements in Stars *Reviews of Modern Physics* **29**, pp. 547–650

(2) Clayton, D. D. (1983). Principles of stellar evolution and nucleosynthesis, *The University of Chicago Press*

(3) Wallerstein, G. *et al* (1997). Synthesis of the elements in stars: forty years of progress *Reviews of Modern Physics* **69**, pp. 995–1084

(4) Woosley, S. E., Heger, A., and Weaver, T. A. (2002). The evolution and explosion of massive stars *Reviews of Modern Physics* **74**, pp. 1015–1071

(5) Clayton, D. D. (2003). Handbook of isotopes in the cosmos : hydrogen to gallium *Cambridge: Cambridge University Press*

C.3   AGB stellar evolution and nucleosynthesis

(1) Iben, I., Jr., and Renzini, A. (1983). Asymptotic giant branch evolution and beyond *Annual review of astronomy and astrophysics* **21**, pp. 271–342

(2) Busso, M., Gallino, R., and Wasserburg, G. J. (1999). Nucleosynthesis in Asymptotic Giant Branch Stars: Relevance for Galactic Enrichment and Solar System Formation *Annual Review of Astronomy and Astrophysics* **37**, pp. 239–309

(3) Habing, H. J., and Olofsson, H. eds. (2003). Asymptotic Giant Branch Stars *New York, Berlin: Springer*

(4) Herwig, F. (2005). Evolution of Asymptotic Giant Branch Stars *Annual Review of Astronomy and Astrophysics* **43**, Planned volume publication date Sep 2005

Bibliography

[1] Abia, C., Domínguez, I., Gallino, R., Busso, M., Masera, S., Straniero, O., de Laverny, P., Plez, B., and Isern, J. (2002). s-Process Nucleosynthesis in Carbon Stars *The Astrophysical Journal* **579**, pp. 817–831

[2] Alexander, C. M. O'd. (1993). Presolar SiC in chondrites - How variable and how many sources? *Geochimica et Cosmochimica Acta* **57**, pp. 2869–2888

[3] Alexander, C. M. O'd., and Nittler, L. R. (1999). The Galactic Evolution of Si, Ti, and O Isotopic Ratios *The Astrophysical Journal* **519**, pp. 222–235

[4] Alexander, C. M. O'd., Swan, P., and Walker, R. M. (1990). In situ measurement of interstellar silicon carbide in two CM chondrite meteorites *Nature* **348**, pp. 715–717

[5] Alpher, R. A., and Herman, R. C. (1950). Theory of the Origin and Relative Abundance Distribution of the Elements *Reviews of Modern Physics* **22**, pp. 153–212

[6] Amari, S., Anders, A., Virag, A., and Zinner, E. (1990). Interstellar graphite in meteorites *Nature* **345**, pp. 238–240

[7] Amari, S. Gao, X., Nittler, L. R., Zinner, E., José, J., Hernanz, M., and Lewis, R. S. (2001). Presolar Grains from Novæ *The Astrophysical Journal* **551**, pp. 1065–1072

[8] Amari, S., Hoppe, P., Zinner, E., and Lewis, R. S. (1992). Interstellar SiC with unusual isotopic compositions – Grains from a supernova? *Astrophysical Journal Letters* **394**, pp. L43–L46

[9] Amari, S., Hoppe, P., Zinner, E., and Lewis, R. S. (1993). The Isotopic Compositions and Stellar Sources of Meteoritic Graphite Grains. *Nature* **365**, pp. 806–809

[10] Amari, S., Hoppe, P., Zinner, E., and Lewis, R. S. (1995). Trace-element concentrations in single circumstellar silicon carbide grains from the Murchison meteorite *Meteoritics* **30**, pp. 679–693

[11] Amari, S., Jennings, C., Nguyen, A., Stadermann, F. J., and Zinner, E. (2002). NanoSIMS Isotopic Analysis of Small Presolar SiC Grains from the Murchison and Indarch Meteorites *Lunar and Planetary Science* **33**, abstract # 1207

[12] Amari, S., Lewis, R. S., and Anders, E. (1994). Interstellar grains in mete-

orites. I - Isolation of SiC, graphite, and diamond; size distributions of SiC and graphite *Geochimica et Cosmochimica Acta* **58**, pp. 459–470

[13] Amari, S., Lewis, R. S., and Anders, E. (1995). Interstellar grains in meteorites: III. Graphite and its noble gases *Geochimica et Cosmochimica Acta* **59**, pp. 1411–1426

[14] Amari, S., Nittler, L. R., Zinner, E., Gallino, R., Lugaro, M., and Lewis, R. S. (2001). Presolar SIC Grains of Type Y: Origin from Low-Metallicity Asymptotic Giant Branch Stars, *The Astrophysical Journal* **546**, pp. 248–266

[15] Amari, S., Nittler, L. R., Zinner, E., Lodders, K., and Lewis, R. S. (2001b). Presolar SiC Grains of Type A and B: Their Isotopic Compositions and Stellar Origins, *The Astrophysical Journal* **559**, pp. 463 – 483

[16] Amari, S., Zinner, E., and Lewis, R. S. (1996). ^{41}Ca in Presolar Graphite of Supernova Origin *Astrophysical Journal Letters* **470**, pp. L101–L104

[17] Amari, S., Zinner, E., and Lewis, R. S. (1999). A Singular Presolar SiC Grain with Extreme ^{29}Si and ^{30}Si Excesses *The Astrophysical Journal Letters* **517**, pp. L59–L62

[18] Amari, S., Zinner, E., and Lewis, R. S. (2000). Isotopic compositions of different presolar SiC size fractions from the Murchison meteorite *Meteoritics & Planetary Science* **35**, pp. 997–1014

[19] Amari, S., Zinner, E., and Lewis, R. S. (2004) Comparison Study of Presolar Graphite Separates KE3 and KFA1 from the Murchison Meteorite *Lunar and Planetary Science* **35**, abstract # 2103

[20] Anders, E., and Grevesse, N. (1989). Abundances of the elements - Meteoritic and solar *Geochimica et Cosmochimica Acta* **53**, pp. 197–214

[21] Anders, E., and Zinner, E. (1993). Interstellar grains in primitive meteorites - Diamond, silicon carbide, and graphite *Meteoritics* **28**, pp. 490–514

[22] Spectral features of presolar diamonds in the laboratory and in carbon star atmospheres Authors: Andersen, A. C., Jørgensen, U. G., Nicolaisen, F. M., Sørensen, P. G., and Glejbøl, K. (1998). *Astronomy and Astrophysics* **330**, pp. 1080–1090

[23] Angulo, C., *et al.* (1999). A compilation of charged-particle induced thermonuclear reaction rates *Nuclear Physics A* **656**, pp. 3–183

[24] Arnett, D., and Bazan, G. (1997). Nucleosynthesis and mixing: unexamined assumptions in Astrophysical Implications of the Laboratory Study of Presolar Materials. Edited by Thomas J. Bernatowicz and Ernst Zinner *American Institute of Physics Conference proceedings 402*, pp. 307–315

[25] Arnould, M., Meynet, G., and Paulus, G. (1997). Wolf-Rayet stars and their nucleosynthesis signatures in meteorites in Astrophysical Implications of the Laboratory Study of Presolar Materials. Edited by Thomas J. Bernatowicz and Ernst Zinner *American Institute of Physics Conference proceedings 402*, pp. 179–202

[26] Bao, Z. Y., Beer, H., Käppeler, F., Voss, F., Wisshak, K., and Rauscher, T. (2000). Neutron Cross Sections for Nucleosynthesis Studies *Atomic Data and Nuclear Data Tables* **76**, pp. 70–154

[27] Becker, S. A., and Iben, I., Jr. (1979). The asymptotic giant branch evo-

lution of intermediate-mass stars as a function of mass and composition. I - Through the second dredge-up phase *Astrophysical Journal* **232**, pp. 831–853

[28] Bernatowicz, T. J., and Zinner, E., eds. (1997). Astrophysical Implications of the Laboratory Study of Presolar Materials *American Institute of Physics Conference Proceedings* **402**

[29] Bernatowicz, T. J., Cowsik, R., Gibbons, P. C., Lodders, K., Fegley, B. Jr., Amari, S., and Lewis, R. S. (1996). Constraints on Stellar Grain Formation from Presolar Graphite in the Murchison Meteorite *Astrophysical Journal* **472**, pp. 760–782

[30] Bernatowicz, T., Fraundorf, G., Ming, T., Anders, E., Wopenka, B., Zinner, E., and Fraundorf, P. (1987). Evidence for interstellar SiC in the Murray carbonaceous meteorite, *Nature* **330**, pp. 728–30

[31] Besmehn, A., and Hoppe, P. (2003). A NanoSIMS study of Si- and Ca-Ti-isotopic compositions of presolar silicon carbide grains from supernovæ *Geochimica et Cosmochimica Acta* **67**, pp. 4693–4703

[32] Bethe, H. A. (1939). Energy Production in Stars, *Physical Review* **55** pp. 434–456

[33] Black, D. C., and Pepin, R. O. (1969). Trapped neon in meteorites – II, *Earth and Planetary Science Letters* **6**, pp. 395-405

[34] Bloecker, T. (1995). Stellar evolution of low and intermediate-mass stars. I. Mass loss on the AGB and its consequences for stellar evolution *Astronomy and Astrophysics* **297**, pp. 727–738

[35] Boothroyd, A. I., and Sackmann, I.-J. (1988a). Low-mass stars. I. Flash-driven luminosity and radius variations *Astrophysical Journal* **328**, pp. 632–640

[36] Boothroyd, A. I., and Sackmann, I.-J. (1988b). Low-mass stars. II. The Core Mass–Luminosity Relations for Low-Mass Stars *Astrophysical Journal* **328**, pp. 641–652

[37] Boothroyd, A. I., and Sackmann, I.-J. (1988c). Low-mass stars. III. Low-Mass Stars with Steady Mass Loss: Up to the Asymptotic Giant Branch and through the Final Thermal Pulses *Astrophysical Journal* **328**, pp. 653–670

[38] Boothroyd, A. I., and Sackmann, I.-J. (1988d). Low-mass stars. IV Carbon stars *Astrophysical Journal* **328**, pp. 671–679

[39] Boothroyd, A. I., and Sackmann, I.-J. (1999). The CNO Isotopes: Deep Circulation in Red Giants and First and Second Dredge-up *The Astrophysical Journal* **510**, pp. 232–250

[40] Boothroyd, A. I., Sackmann, I.-J., and Ahern, S. C. (1993). Prevention of High-Luminosity Carbon Stars by Hot Bottom Burning *Astrophysical Journal* **416**, pp. 762–768

[41] Boothroyd, A. I., Sackmann, I.-J., and Wasserburg, G. J. (1994). Predictions of oxygen isotope ratios in stars and of oxygen-rich interstellar grains in meteorites *The Astrophysical Journal Letters* **430**, pp. L77–L80

[42] Boothroyd, A. I., Sackmann, I.-J., and Wasserburg, G. J. (1995). Hot bottom burning in asymptotic giant branch stars and its effect on oxygen isotopic abundances Authors: *Astrophysical Journal* **442**, pp. L21–L24

[43] Boss, A. P., and Vanhala, H. A. T. (2002). Triggering Protostellar Collapse, Injection, and Disk Formation *Space Science Reviews* **92**, pp. 13–22

[44] Bouchet, P., De Buizer, J. M., Suntzeff, N. B., Danziger, I. J., Hayward, T. L., Telesco, C. M., and Packham, C. (2004). High-Resolution Mid-infrared Imaging of SN 1987A *The Astrophysical Journal* **611**, pp. 394–398

[45] Bowen, G. H. (1988). Dynamical modeling of long-period variable star atmospheres *Astrophysical Journal* **329**, pp. 299–317

[46] Brown, L. E., and Clayton, D. D. (1992). Silicon isotopic composition in large meteoritic SiC particles and ^{22}Na origin of ^{22}Ne *Science* **258**, pp. 970–972

[47] Burbidge, E. M., Burbidge, G. R., Fowler, W. A., and Hoyle, F. (1957). Synthesis of the Elements in Stars *Reviews of Modern Physics* **29**, pp. 547–650

[48] Busso, M., Gallino, R., Lambert, D. L., Travaglio, C., and Smith, V. V. (2001). Nucleosynthesis and Mixing on the Asymptotic Giant Branch. III. Predicted and Observed s-Process Abundances *The Astrophysical Journal* **557**, pp. 802–821

[49] Busso, M., Gallino, R., and Wasserburg, G. J. (1999). Nucleosynthesis in Asymptotic Giant Branch Stars: Relevance for Galactic Enrichment and Solar System Formation *Annual Review of Astronomy and Astrophysics* **37**, pp. 239–309

[50] Cameron, A. G. W. (1955). Origin of Anomalous Abundances of the Elements in Giant Stars *Astrophysical Journal* **121**, pp.144–160

[51] Cameron, A. G. W., and Fowler, W. A. (1971). Lithium and the s-process in Red-Giant Stars *Astrophysical Journal* **164**, pp. 111–114

[52] Cameron, A. G. W. (2001). Some Properties of r-Process Accretion Disks and Jets *The Astrophysical Journal* **562**, pp. 456–469

[53] Cameron, A. G. W., and Truran, J. W. (1977). The supernova trigger for formation of the solar system *Icarus* **30**, pp. 447–461

[54] Cassen, P., and Chick, K. M. (1997). The Survival of Presolar Grains during the Formation of the Solar System, in Astrophysical Implications of the Laboratory Study of Presolar Materials. Edited by Thomas J. Bernatowicz and Ernst Zinner *American Institute of Physics Conference proceedings 402*, pp. 697–791

[55] Charbonnel, C. (1994). Clues for non-standard mixing on the red giant branch from ^{12}C/^{13}C and ^{12}C/^{14}N ratios in evolved stars *Astronomy and Astrophysics* **282**, pp. 811–820

[56] Charbonnel, C., and Balachandran, S. C. (2000). The Nature of the lithium rich giants. Mixing episodes on the RGB and early-AGB *Astronomy and Astrophysics* **359**, pp. 563–572

[57] Cherchneff, I., and Cau, P. (1999). The chemistry of carbon dust formation in Asymptotic Giant Branch Stars, IAU Symposium #191, Edited by T. Le Bertre, A. Lebre, and C. Waelkens, pp. 251–259

[58] Choi, B., Huss, G. R., Wasserburg, G. J., and Gallino, R. (1998). Presolar Corundum and Spinel in Ordinary Chondrites: Origins from AGB Stars and a Supernova *Science* **282**, pp. 1284–1289

[59] Choi, B., Wasserburg, G. J., and Huss, G. R. (1999). Circumstellar Hibonite and Corundum and Nucleosynthesis in Asymptotic Giant Branch Stars *The Astrophysical Journal* **522**, pp. L133-L136

[60] Clayton, D. D. (1983). Principles of stellar evolution and nucleosynthesis, *The University of Chicago Press*

[61] Clayton, D. D. (1975). ^{22}Na, Ne-E, extinct radioactive anomalies and unsupported ^{40}Ar *Nature* **257**, pp. 36–37

[62] Clayton, D. D. (1983) Discovery of s-process Nd in Allende residue *Astrophysical Journal Letters* **271**, pp. L107–L109

[63] Clayton, D. D. (1989). Origin of heavy xenon in meteoritic diamonds *Astrophysical Journal* **340**, pp. 613–619

[64] Clayton, D. D. (1997). Placing the Sun and Mainstream SiC Particles in Galactic Chemodynamic Evolution *Astrophysical Journal Letters* **484**, pp. L67–70

[65] Clayton, D. D. (2003). A Presolar Galactic Merger Spawned the SiC-Grain Mainstream *The Astrophysical Journal* **598**, pp. 313–324

[66] Clayton, D. D., Arnett, D., Kane, J., and Meyer, B. S. (1997). Type X Silicon Carbide Presolar Grains: Type Ia Supernova Condensates? *Astrophysical Journal* **486**, pp. 824–834

[67] Clayton, D. D., Deneault, E. A.-N., and Meyer, B. S. (2001). Condensation of Carbon in Radioactive Supernova Gas *The Astrophysical Journal* **562**, pp. 480–493

[68] Clayton, D. D., and Nittler, L. R. (2004). Astrophysics with Presolar Stardust *Annual Review of Astronomy and Astrophysics* **42**, pp. 39–78

[69] Clayton, D. D., Liu, W., and Dalgarno, A. (1999). Condensation of carbon in radioactive supernova gas *Science* **283**, pp. 1290–1292

[70] Clayton, D. D., and Timmes, F. X. (1997). Placing the Sun in Galactic Chemical Evolution: Mainstream SiC Particles *Astrophysical Journal* **483**, p.220–227

[71] Clayton, D. D., and Ward, R. A. (1978). S-process studies - Xenon and krypton isotopic abundances *Astrophysical Journal* **224**, pp. 1000–1006

[72] Clayton, R. N., Grossman, L., and Mayeda, T. K. (1973). A component of primitive nuclear composition in carbonaceous meteorites *Science* **182**, pp. 485–488

[73] Clayton, R. N., Hinton, R. W., and Davis, A. M. (1988). Isotopic variations in the rock-forming elements in meteorites *Royal Society (London), Philosophical Transactions, Series A* **325**, pp. 483–501.

[74] Clément, D., Mutschke, H., Klein, R., and Henning, Th. (2003). New Laboratory Spectra of Isolated β-SiC Nanoparticles: Comparison with Spectra Taken by the Infrared Space Observatory *The Astrophysical Journal* **594**, pp. 642–650

[75] Croat, T. K., Bernatowicz, T., Amari, S., Messenger, S., and Stadermann, F. J. (2003). Structural, chemical, and isotopic microanalytical investigations of graphite from supernovæ *Geochimica et Cosmochimica Acta* **67**, pp. 4705–4725

[76] Dai, Z. R., Bradley, J. P., Joswiak, D. J., Brownlee, D. E., Hill, H. G. M., and Genge, M. J. (2002). Possible in situ formation of meteoritic nanodiamonds in the early Solar System *Nature* **418**, pp. 157–159

[77] Daulton, T. L., Bernatowicz, T. J., Lewis, R. S., Messenger, S., Stader-

mann, F. J., and Amari, S. (2002). Polytype distribution in circumstellar silicon carbide *Science* **296**, pp. 1852–1855

[78] Daulton, T. L., Eisenhour, D. D., Bernatowicz, T. J., Lewis, R. S., and Buseck, P. R. (1996). Genesis of presolar diamonds: Comparative high-resolution transmission electron microscopy study of meteoritic and terrestrial nano-diamonds *Geochimica et Cosmochimica Acta* **60**, pp. 4853–4872

[79] Dearborn, D. S. P. (1992). Diagnostics of stellar evolution: The oxygen isotopes *Physics Reports* **210**, pp. 367–382

[80] Demyk, K., Dartois, E., Wiesemeyer, H., Jones, A. P., and d'Hendecourt, L. (2000) Structure and chemical composition of the silicate dust around OH/IR stars *Astronomy and Astrophysics* **364**, pp. 170–178

[81] Deneault, E. A.-N., Clayton, D. D., and Heger, A. (2003). Supernova Reverse Shocks: SiC Growth and Isotopic Composition *The Astrophysical Journal* **594**, pp. 312–325.

[82] Denissenkov, P. A., and Tout, C. A. (2003). Partial mixing and formation of the ^{13}C pocket by internal gravity waves in asymptotic giant branch stars *Monthly Notice of the Royal Astronomical Society* **340**, pp. 722–732

[83] Denissenkov, P. A., and VandenBerg, D. A. (2003). Canonical Extra Mixing in Low-Mass Red Giants *The Astrophysical Journal* **593**, pp. 509–523

[84] Desch, S. J., Connolly, H. C., Jr., and Srinivasan, G. (2004). An Interstellar Origin for the Beryllium 10 in Calcium-rich, Aluminum-rich Inclusions *Astrophysical Journal* **602**, pp. 528–542

[85] Despain, K. H. (1980). A difficulty with Ne-22 as the neutron source for the solar system s-process Authors: *Astrophysical Journal Letters* **236**, pp. L165–L168

[86] Dunbar, D. N. F., Pixley, R. E., Wenzel, W. A., and Whaling, W. (1953). The 7.68-Mev State in ^{12}C *Physical Review* **92**, pp. 649–650

[87] Dunner, L., Eales, S., Ivison, R., Morgan, H., and Edmunds, M. (2003). Type II supernovæ as a significant source of interstellar dust *Nature* **424**, pp. 285–287

[88] Ebel, D. S., and Grossman, L. (2001). Condensation from supernova gas made of free atoms *Geochimica et Cosmochimica Acta* **65**, pp. 469–477

[89] El Eid, M. F. (1994). CNO isotopes in red giants: theory versus observations *Astronomy and Astrophysics* **285**, pp. 915–928

[90] Feast, M. W. (1989). The kinematics of peculiar red giants, in IAU Colloq. #106, Evolution of Peculiar Red Giant Stars, ed. H. R. Johnson & B. Zuckerman (Cambridge: Cambridge Univ. Press), pp. 26–34

[91] Forestini, M., Arnould, M., and Paulus, G. (1991). On the production of ^{91}Al in AGB stars *Astronomy and Astrophysics* **252**, pp. 597–604

[92] Forestini, M., and Charbonnel, C. (1997). Nucleosynthesis of light elements inside thermally pulsing AGB stars: I. The case of intermediate-mass stars *Astronomy and Astrophysics Supplement series* **123**, pp. 241–272

[93] Freytag, B., Ludwig, H.-G., and Steffen, M. (1996). Hydrodynamical models of stellar convection. The role of overshoot in DA white dwarfs, A-type stars, and the Sun *Astronomy and Astrophysics* **313**, pp. 497–516

[94] Frogel, J. A., Mould, J., and Blanco, V. M. (1990). The asymptotic giant branch of Magellanic Cloud clusters *Astrophysical Journal* **352**, pp. 96–122

[95] Frost, C. A., and Lattanzio, J. C. (1996). On the Numerical Treatment and Dependence of the Third Dredge-up Phenomenon *Astrophysical Journal* **473**, pp. 383–387

[96] Fujimoto, S., Hashimoto, M., Koike, O., Arai, K., and Matsuba, R. (2003). *p*-Process Nucleosynthesis inside Supernova-driven Supercritical Accretion Disks *The Astrophysical Journal* **585**, pp. 418–428

[97] Gallino, R., Arlandini, C., Busso, M., Lugaro, M., Travaglio, C., Straniero, O., Chieffi, A., and Limongi, M. (1998). Evolution and Nucleosynthesis in Low-Mass Asymptotic Giant Branch Stars. II. Neutron Capture and the s-Process *Astrophysical Journal* **497**, pp. 388–403

[98] Gallino, R., Busso, M., and Lugaro, M. (1997). Neutron Capture Nucleosynthesis in AGB Stars, in Astrophysical Implications of the Laboratory Study of Presolar Materials. Edited by Thomas J. Bernatowicz and Ernst Zinner *American Institute of Physics Conference Proceedings 402*, pp. 115–153

[99] Gallino, R., Busso, M., Picchio, G., and Raiteri, C. M. (1990). On the astrophysical interpretation of isotope anomalies in meteoritic SiC grains *Nature* **348**, pp. 298–302

[100] Gallino, R., Raiteri, C. M., and Busso, M. (1993). Carbon stars and isotopic Ba anomalies in meteoritic SiC grains *Astrophysical Journal* **410**, pp. 400–411

[101] Gallino, R., Raiteri, C. M., Busso, M., and Matteucci, F. (1994). The puzzle of silicon, titanium, and magnesium anomalies in meteoritic silicon carbide grains *The Astrophysical Journal* **430**, pp. 858–869

[102] Gilroy, K. K. (1989). Carbon isotope ratios and lithium abundances in open cluster giants *Astrophysical Journal* **347**, pp. 835–848

[103] Goriely, S., José, J., Hernanz, M., Rayet, M., and Arnould, M. (2002). He-detonation in sub-Chandrasekhar CO white dwarfs: A new insight into energetics and p-process nucleosynthesis *Astronomy and Astrophysics Letters* **383**, pp. L27–L30

[104] Goriely, S., and Mowlavi, N. (2000). Neutron-capture nucleosynthesis in AGB stars *Astronomy and Astrophysics* **362**, pp. 599–614

[105] Greenstein, J. L. (1956). The Abundances of the Chemical Elements in the Galaxy and the Theory of Their Origin *Publications of the Astronomical Society of the Pacific* **68**, pp. 185–203

[106] Groenewegen, M. A. T., van den Hoek, L. B., and de Jong, T. (1995). The evolution of galactic carbon stars. *Astronomy & Astrophysics* **293**,pp. 381–395

[107] Guelin, M., Forestini, M., Valiron, P., Ziurys, L. M., Anderson, M. A., Cernicharo, J., and Kahane, C. (1995). Nucleosynthesis in AGB stars: Observation of ^{25}Mg and ^{26}Mg-26 in IRC+10216 and possible detection of ^{26}Al *Astronomy and Astrophysics* **297**, pp. 183–196

[108] Haas, M. R., Erickson, E. F., Lord, S. D., Hollenbach, D. J., Colgan, S. W. J., and Burton, M. G. (1990). Velocity-resolved far-infrared spectra of forbidden Fe II - Evidence for mixing and clumping in SN 1987A *Astrophysical Journal* **360**, pp. 257–266

[109] Habing, H.J., and Olofsson, H. eds. (2003). Asymptotic Giant Branch Stars *New York, Berlin: Springer*

[110] Harris, M. J., Lambert, D. L., and Smith, V. V. (1985). Oxygen isotopic abundances in evolved stars. I - Six barium stars *Astrophysical Journal* **292**, pp. 620–627

[111] Harris, M. J., Lambert, D. L., Hinkle, K. H., Gustafsson, B., and Eriksson, K. (1987). Oxygen isotopic abundances in evolved stars. III - 26 carbon stars *Astrophysical Journal* **316**, pp. 294–304

[112] Harris, M. J., Lambert, D. L., and Smith, V. V. (1988). Oxygen isotopic abundances in evolved stars. IV - Five K giants *Astrophysical Journal* **325**, pp. 768–775

[113] Haseltine, E. (2002). The 11 Greatest Unanswered Questions of Physics *Discover magazine* **23**, No. 2

[114] Heinrich, M., Huss, G. R., and Wasserburg, G. J. (1998). Automated SEM Identification and Location of Rare Phases *Lunar and Planetary Science* **29**, abstract # 1715

[115] Herwig, F. (2000). The evolution of AGB stars with convective overshoot *Astronomy and Astrophysics* **360**, pp. 952–968

[116] Herwig, F. (2005). Evolution of Asymptotic Giant Branch Stars *Annual Review of Astronomy and Astrophysics* **43**, Planned volume publication date September 2005

[117] Herwig, F., Bloecker, T., Schoenberner, D., and El Eid, M. (1997). Stellar evolution of low and intermediate-mass stars. IV. Hydrodynamically-based overshoot and nucleosynthesis in AGB stars. *Astronomy and Astrophysics* **324**, pp. L81–L84

[118] Herwig, F., Langer, N., and Lugaro, M. (2003). The s-Process in Rotating Asymptotic Giant Branch Stars *The Astrophysical Journal* **593**, pp. 1056–1073

[119] Heymann, D., and Dziczkaniec, M. (1980a). Xenon, osmium, and lead formed in O-shells and C-shells of massive stars *Meteoritics* **15**, pp. 1–14

[120] Heymann, D., and Dziczkaniec, M. (1980b). A process of stellar nucleosynthesis which mimicks mass fractionation in P-xenon *Meteoritics* **15**, 1980, pp. 15–24

[121] Hoppe, P. (2002). NanoSIMS perspectives for nuclear astrophysics *New Astronomy Reviews* **46**, pp. 589–595

[122] Hoppe, P., Amari, S., Zinner, E., Ireland, T., and Lewis, R. S. (1994). Carbon, nitrogen, magnesium, silicon, and titanium isotopic compositions of single interstellar silicon carbide grains from the Murchison carbonaceous chondrite *The Astrophysical Journal* **430**, pp. 870–890

[123] Hoppe, P., Amari, S., Zinner, E., and Lewis, R. S. (1995). Isotopic compositions of C, N, O, Mg, and Si, trace element abundances, and morphologies of single circumstellar graphite grains in four density fractions from the Murchison meteorite *Geochimica et Cosmochimica Acta* **59**, pp. 4029–4056

[124] Hoppe, P., Annen, P., Strebel, R., Eberhardt, P., Gallino, R., Lugaro, M., Amari, S., and Lewis, R. S. (1997). Meteoritic Silicon Carbide Grains with Unusual Si Isotopic Compositions: Evidence for an Origin in Low-Mass,

Low-Metallicity Asymptotic Giant Branch Stars *Astrophysical Journal Letters* **487**, pp. L101–104

[125] Hoppe, P., and Besmehn, A. (2002). Evidence for Extinct Vanadium-49 in Presolar Silicon Carbide Grains from Supernovæ *The Astrophysical Journal* **576**, pp. L69–L72

[126] Hoppe, P., and Ott, U. (1997). Mainstream silicon carbide grains from meteorites, in Astrophysical Implications of the Laboratory Study of Presolar Materials. Edited by Thomas J. Bernatowicz and Ernst Zinner *American Institute of Physics Conference proceedings 402*, pp. 27–58

[127] Hoppe, P., Strebel, R., Eberhardt, P., Amari, S., and Lewis, R. S. (1996). Small SiC grains and a nitride grain of circumstellar origin from the Murchison meteorite: Implications for stellar evolution and nucleosynthesis *Geochimica et Cosmochimica Acta* **60**, pp. 883–907

[128] Type II Supernova Matter in a Silicon Carbide Grain from the Murchison Meteorite Hoppe, P., Strebel, R., Eberhardt, P., Amari, S., and Lewis, R. S. (1996) *Science* **272**, pp. 1314–1316

[129] Hoppe, P., Strebel, R., Eberhardt, P., Amari, S., and Lewis, R. S. (2000). Isotopic properties of silicon carbide X grains from the Murchison meteorite in the size range 0.5-1.5 μm *Meteoritics & Planetary Science* **35**, pp. 1157–1176

[130] Howard, W. M., Meyer, B. S., and Clayton, D. D. (1992). Heavy-element abundances from a neutron burst that produces Xe-H *Meteoritics* **27**, pp. 404–412

[131] Hoyle, F. (1946). The synthesis of the elements from hydrogen, *Monthly Notices of the Royal Astronomical Society* **106**, pp.343–383

[132] Huss, G. R. (1997). The Survival of Presolar Grains in Solar System Bodies, in Astrophysical Implications of the Laboratory Study of Presolar Materials. Edited by Thomas J. Bernatowicz and Ernst Zinner *American Institute of Physics Conference proceedings 402*, pp. 721–748

[133] Huss, G. R., Fahey, A. J., Gallino, R., and Wasserburg, G. J. (1994). Oxygen isotopes in circumstellar Al_2O_3 grains from meteorites and stellar nucleosynthesis *The Astrophysical Journal Letters* **430**, pp. L81–L84

[134] Huss, G. R., and Lewis, R. S. (1994). Noble gases in presolar diamonds I: Three distinct components and their implications for diamond origins *Meteoritics* **29**, pp. 791–810

[135] Huss, G. R., Hutcheon, I. D., and Wasserburg, G. J. (1997). Isotopic systematics of presolar silicon carbide from the Orgueil (CI) chondrite: Implications for solar system formation and stellar nucleosynthesis *Geochimica et Cosmochimica Acta* **61**, pp. 5117–5148

[136] Hutcheon, I. D., Huss, G. R., Fahey, A. J., and Wasserburg, G. J. (1994). Extreme Mg-26 and O-17 enrichments in an Orgueil corundum: Identification of a presolar oxide grain *Astrophysical Journal Letters* **425**, pp. L97–L100

[137] Iben, I., Jr. (1975). Thermal pulses; p-capture, alpha-capture, s-process nucleosynthesis; and convective mixing in a star of intermediate mass *Astrophysical Journal* **196**, pp. 525–547

[138] Iben, I., Jr. (1975). Neon-22 as a neutron source, light elements as modulators, and s-process nucleosynthesis in a thermally pulsing star *Astrophysical Journal* **196** pp. 549–558

[139] Iben, I., Jr. (1976). Further adventures of a thermally pulsing star *Astrophysical Journal* **208**, pp. 165–176

[140] Iben, I., Jr. (1985). The life and times of an intermediate mass star - In isolation/in a close binary *Royal Astronomical Society, Quarterly Journal* **26**, pp. 1–39

[141] Iben, I., Jr., and Renzini, A. (1982). On the formation of carbon star characteristics and the production of neutron-rich isotopes in asymptotic giant branch stars of small core mass *Astrophysical Journal Letters* **263**, pp. L23–L27

[142] Iben, I., Jr., and Renzini, A. (1983). Asymptotic giant branch evolution and beyond *Annual review of astronomy and astrophysics* **21**, pp. 271–342

[143] Imbriani, G., Limongi, M., Gialanella, L., Terrasi, F., Straniero, O., and Chieffi, A. (2001). The $^{12}C(\alpha,\gamma)^{16}O$ Reaction Rate and the Evolution of Stars in the Mass Range $0.8 \leq M/M_\odot \leq 25$ *The Astrophysical Journal* **558**, pp. 903 – 915

[144] Iwamoto, N., Kajino, T., Mathews, G. J., Fujimoto, M. Y., and Aoki, W. (2004). Flash-Driven Convective Mixing in Low-Mass, Metal-deficient Asymptotic Giant Branch Stars: A New Paradigm for Lithium Enrichment and a Possible s-Process *The Astrophysical Journal* **602**, pp. 377–388

[145] Jaeger, M., Kunz, R., Mayer, A., Hammer, J. W., Staudt, G., Kratz, K. L., and Pfeiffer, B. (2001). $^{22}Ne(\alpha,n)^{25}Mg$: The Key Neutron Source in Massive Stars *Physical Review Letters* **87**, id. 202501

[146] Jennings, C. L., Savina, M. R., Messenger, S., Amari, S., Nichols, R. H., Jr., Pellin, M. J., and Podosek, F. A. (2002). Indarch SiC by TIMS, RIMS, and NanoSIMS *Lunar and Planetary Science* **33**, abstract # 1833

[147] Jones, A. P, Tielens, A. G. G. M., Hollenbach, D. J., and McKee, C. F. (1997). The Propagation and Survival of Interstellar Grains, in Astrophysical Implications of the Laboratory Study of Presolar Materials. Edited by Thomas J. Bernatowicz and Ernst Zinner *American Institute of Physics Conference proceedings 402*, pp. 595–611

[148] Jørgensen, U. G. (1988). Formation of Xe-HL-enriched diamond grains in stellar environments *Nature* **332**, pp. 702–705

[149] Jorissen, A., and Goriely, S. (2001). The Nuclear Network Generator: A tool for nuclear astrophysics *Nuclear Physics A* **688**, pp. 508–510

[150] Jura, M. (1997). carbon dust particle size distribution around mass-losing AGB stars, in Astrophysical Implications of the Laboratory Study of Presolar Materials. Edited by Thomas J. Bernatowicz and Ernst Zinner *American Institute of Physics Conference proceedings 402*, pp. 379–390.

[151] Jura, M., Webb, R. A., and Kahane, C. (2001). Large Circumbinary Dust Grains around Evolved Giants? *The Astrophysical Journal Letters* **550**, pp. L71–L75.

[152] Kahane, C., Cernicharo, J., Gomez-Gonzalez, J., and Guelin, M. (1992). Isotopic abundances in carbon-rich circumstellar envelopes - A further it-

eration on the oxygen isotope puzzle *Astronomy and Astrophysics* **256**, pp. 235–250

[153] Kahane, C., Gomez-Gonzalez, J., Cernicharo, J., and Guelin, M. (1988). Carbon, nitrogen, sulfur and silicon isotopic ratios in the envelope of IRC + 10216 *Astronomy and Astrophysics* **190**, pp. 167–177

[154] Käppeler, F., Wiescher, M., Giesen, U., Goerres, J., Baraffe, I., El Eid, M., Raiteri, C. M., Busso, M., Gallino, R., Limongi, M., and Chieffi, A. (1994) Reaction rates for O-18(alpha, gamma)Ne-22, Ne-22(alpha, gamma)Mg-26, and Ne-22(alpha, n)Mg-25 in stellar helium burning and s-process nucleosynthesis in massive stars *Astrophysical Journal* **437**, pp. 396–409

[155] Kashiv, Y., Cai, Z., Lai, B., Sutton, S. R., Lewis, R. S., Davis, A. M., Clayton, R. N., and Pellin, M. J. (2001). Synchrotron X-Ray Fluorescence: A New Approach for Determining Trace Element Concentrations in Individual Presolar SiC Grains *Lunar and Planetary Science* **32**, abstract # 2192

[156] Kashiv, Y., Cai, Z., Lai, B., Sutton, S. R., Lewis, R. S., Davis, A. M., Clayton, R. N., and Pellin, M. J. (2002). Condensation of Trace Elements into Presolar SiC Stardust Grains *Lunar and Planetary Science* **33**, abstract # 2056

[157] Kehm, K., Amari, S., Hohenberg, C. M., and Lewis, R. S. (1996). 22Ne-E(L) Measured in Individual KFC1 Graphite Grains from the Murchison Meteorite *Lunar and Planetary Science* **27**, pp. 657–658

[158] Kratz, K.-L., Bitouzet, J.-P., Thielemann, F.-K., Moeller, P., and Pfeiffer, B. (1993). Isotopic r-process abundances and nuclear structure far from stability — Implications for the r-process mechanism *Astrophysical Journal* **403**, pp. 216–238.

[159] Lamb, S. A., Howard, W. M., Truran, J. W., and Iben, I., Jr. (1977). Neutron-capture nucleosynthesis in the helium-burning cores of massive stars *Astrophysical Journal* **217**, pp. 213–221

[160] Lambert, D. L., Gustafsson, B., Eriksson, K., and Hinkle, K. H. (1986). The chemical composition of carbon stars. I - Carbon, nitrogen, and oxygen in 30 cool carbon stars in the Galactic disk *Astrophysical Journal Supplement Series* **62**, pp. 373–425

[161] Langer, N., Heger, A., Wellstein, S., and Herwig, F. (1999). Mixing and nucleosynthesis in rotating TP-AGB stars *Astronomy and Astrophysics* **346**, pp. L37–L40

[162] Lattanzio, J. C. (1986). The asymptotic giant branch evolution of 1.0-3.0 solar mass stars as a function of mass and composition *Astrophysical Journal* **311**, pp. 708–730

[163] Lattanzio, J. C. (1987). The formation of a 1.5-solar mass carbon star with Mbol = -4.4 *Astrophysical Journal Letters* **313**, pp. L15–L18

[164] Lattanzio, J. C. (1989). Carbon dredge-up in low-mass stars and solar metallicity stars *Astrophysical Journal Letters* **344**, pp. L25–L27

[165] Lee, T., Papanastassiou, D. A., and Wasserburg, G. J. (1977). Aluminum-26 in the early solar system - Fossil or fuel *Astrophysical Journal Letters* **211**, pp. L107–L110

[166] Lee, T., Shu, F. H., Shang, H., Glassgold, A. E., and Rehm, K. E. (1998). Protostellar Cosmic Rays and Extinct Radioactivities in Meteorites *Astrophysical Journal* **506**, pp. 898–912

[167] Lewis, R. S., Amari, S., and Anders, E. (1990). Meteoritic silicon carbide - Pristine material from carbon stars *Nature* **348**, pp. 293–298

[168] Lewis, R. S., Amari, S., and Anders, E. (1994). Interstellar grains in meteorites: II. SiC and its noble gases *Geochimica et Cosmochimica Acta* **58**, pp. 471–494

[169] Lewis, R. S., Tang, M., Wacker, J. F., Anders, E., and Steel, E. (1987). Interstellar diamonds in meteorites, *Nature* **326**, pp. 160–62

[170] Lin, Y., Amari, S., and Pravdivtseva, O. (2002). Presolar Grains from the Qingzhen (EH3) Meteorite *The Astrophysical Journal* **575**, pp. 257–263

[171] Lodders, K., and Amari, S. (2005). Presolar grains from meteorites: Remnants from the early times of the solar system *Chemie der Erde*, to appear

[172] Lodders, K., and Fegley, B. Jr. (1995). The origin of circumstellar silicon carbide grains found in meteorites *Meteoritics* **30**, pp. 661–678

[173] Lodders, K., and Fegley, B. Jr. (1997). Complementary Trace Element Abundances in Meteoritic SiC Grains and Carbon Star Atmospheres *Astrophysical Journal Letters* **484**, pp. L71–74

[174] Lugaro, M., Davis, A. M., Gallino, R., Pellin, M. J., Straniero, O., and Käppeler, F. (2003). Isotopic Compositions of Strontium, Zirconium, Molybdenum, and Barium in Single Presolar SiC Grains and Asymptotic Giant Branch Stars Authors: *The Astrophysical Journal* **593**, pp. 486–508

[175] Lugaro, M., Herwig, F., Lattanzio, J. C., Gallino, R., and Straniero, O. (2003). s-Process Nucleosynthesis in Asymptotic Giant Branch Stars: A Test for Stellar Evolution *The Astrophysical Journal* **586**, pp. 1305–1319

[176] Lugaro, M., Pols, O., Karakas, A., and Tout, C. (2005). HR4049: Signature of Nova Nucleosynthesis? *Nuclear Physics A*, to appear

[177] Lugaro, M., Zinner, E., Gallino, R., and Amari, S. (1999). Si Isotopic Ratios in Mainstream Presolar SiC Grains Revisited *The Astrophysical Journal* **527**, pp. 369–394

[178] Ma, Z., Thompson, R. N., Lykke, K. R., Pellin, M. J., and Davis, A. M. (1995). New instrument for microbeam analysis incorporating submicron imaging and resonance ionization mass spectrometry. *Review of Scientific Instruments* **66**, pp. 3168–3176

[179] Maas, R., Loss, R. D., Rosman, K. J. R., de Laeter, J. R., Lewis, R. S., Huss, G. R., and Lugmair, G. W. (2001). Isotope anomalies in tellurium and palladium from Allende nanodiamonds Authors: *Meteoritics & Planetary Science* **36**, pp. 849–858

[180] Marhas, K. K., Hoppe, P., and Besmehn, A. (2004). A NanoSIMS Study of Iron-Isotopic Compositions in Presolar Silicon Carbide Grains *Lunar and Planetary Science* **35**, abstract # 1834

[181] Marhas, K. K., Hoppe, P., and Ott, U. (2003). A NanoSIMS Study of C-, Si- and Ba-Isotopic Compositions of Presolar Silicon Carbide Grains from the Murchison Meteorite *Meteoritics & Planetary Science* **38**, abstract # 5101

[182] McWilliam, A., and Lambert, D. L. (1988). Isotopic magnesium abundances in stars *Monthly Notices of the Royal Astronomical Society* **230**, pp. 573–585

[183] Mendybaev, R. A., Beckett, J. R., Grossman, L., Stolper, E., Cooper, R. F., and Bradley, J. (2002). Volatilization kinetics of silicon carbide in reducing gases: an experimental study with applications to the survival of presolar grains in the solar nebula *Geochimica et Cosmochimica Acta* **66**, pp. 661–682

[184] Merrill, P. W. (1952a). Technetium in stars *Science* **115**, p. 484

[185] Merrill, P. W. (1952b). Spectroscopic Observations of Stars of Class S *Astrophysical Journal* **116**, pp.21–26

[186] Messenger, S. (2000). Identification of molecular-cloud material in interplanetary dust particles *Nature* **404**, pp. 968–971

[187] Messenger, S., Amari, S., Gao, X., Walker, R. M., Clemett, S. J., Chillier, X. D. F., Zare, R. N., and Lewis, R. S. (1998). Indigenous Polycyclic Aromatic Hydrocarbons in Circumstellar Graphite Grains from Primitive Meteorites *Astrophysical Journal* **502**, pp. 284–

[188] Messenger, S., Keller, L. P., Stadermann, F. J., Walker, R. M., and Zinner, E. (2003). Samples of Stars Beyond the Solar System: Silicate Grains in Interplanetary Dust *Science* **300**, pp. 105–108

[189] Messenger, S., and Walker, R. M. (1997). Evidence for Molecular Cloud Material in Meteorites and Interplanetary Dust, in Astrophysical Implications of the Laboratory Study of Presolar Materials. Edited by Thomas J. Bernatowicz and Ernst Zinner *American Institute of Physics Conference proceedings 402*, pp. 545–564

[190] Meyer, B. S. (1994). The *r*-, *s*-, and *p*-Processes in Nucleosynthesis *Annual Review of Astronomy and Astrophysics* **32**, pp. 153–190

[191] Meyer, B. S. (2002). *r*-Process Nucleosynthesis without Excess Neutrons *Physical Review Letters* **89**, id. 231101

[192] Meyer, B. S., and Clayton, D. D. (2000). Short-Lived Radioactivities and the Birth of the sun *Space Science Reviews* **92**, pp. 133–152

[193] Meyer, B. S., Clayton, D. D., The, L.-S. (2000). Molybdenum and Zirconium Isotopes from a Supernova Neutron Burst *The Astrophysical Journal* **540**, pp. L49–L52

[194] Mostefaoui, S., and Hoppe, P. (2004). Discovery of Abundant In Situ Silicate and Spinel Grains from Red Giant Stars in a Primtive Meteorite *The Astrophysical Journal* **613**, pp. 149–152

[195] Mostefaoui, S., Hoppe, P., Marhas, K. K., and Gröner, E. (2003). Search for In Situ Presolar Oxygen-rich Dust in Meteorites *Meteoritics & Planetary Science* **38**, abstract # 5185

[196] Mostefaoui, S., Marhas, K. K., and Hoppe, P. (2004). Discovery of an In-Situ Presolar Silicate Grain with GEMS-Like Composition in the Bishunpur Matrix *Lunar Planetary Science* **35**, abstract # 1593

[197] Mowlavi, N. (1999). On the third dredge-up phenomenon in asymptotic giant branch stars *Astronomy and Astrophysics* **344**, pp. 617–631

[198] Mowlavi, N., and Meynet, G. (2000). Aluminum-26 production in asymptotic giant branch stars *Astronomy and Astrophysics* **361**, pp. 959–976

[199] Nagashima, K., Krot, A. N., and Yurimoto, H. (2004). Stardust silicates from primitive meteorites *Nature* **428**, pp. 921–924

[200] Nguyen A. N., and Zinner E. (2004). Discovery of ancient silicate stardust in a meteorite *Science* **303**, pp. 1496–1499

[201] Nichols, R. H., Jr., Hohenberg, C. M., Amari, S., and Lewis, R. S. (1991). ^{22}Ne-E(H) and ^{4}He Measured in individual SiC grains using laser gas extraction *Meteoritics* **26**, pp. 377–378

[202] Nichols, R. H., Jr., Kehm, K., Hohenberg, C. M., Amari, S., and Lewis, R. S. (2005). Neon and Helium in single interstellar SiC and graphite grains: asymptotic giant branch, Wolf-Rayet, supernova and nova sources *Geochimica Cosmochimica Acta*, to appear

[203] Nicolussi, G. K., Davis, A. M., Pellin, M. J., Lewis, R. S., Clayton, R. N., and Amari, S. (1997). *s*-Process zirconium in presolar silicon carbide grains. *Science* **277**, pp. 1281–1283

[204] Nicolussi, G. K., Pellin, M. J., Lewis, R. S., Davis, A. M., Amari, S., and Clayton, R. N. (1998). Molybdenum isotopic composition of individual presolar silicon carbide grains from the Murchison meteorite *Geochimica Cosmochimica Acta* **62**, pp. 1093–1104

[205] Nicolussi, G. K., Pellin, M. J., Lewis, R. S., Davis, A. M., Clayton, R. N., and Amari, S. (1998). Strontium isotopic composition in individual circumstellar silicon carbide grains: A record of s-process nucleosynthesis *Physical Review Letters* **81**, pp. 3583–3586

[206] Nicolussi, G. K., Pellin, M. J., Lewis, R. S., Davis, A. M., Clayton, R. N., and Amari, S. (1998). Zirconium and molybdenum in individual circumstellar graphite grains: New isotopic data on the nucleosynthesis of heavy elements *The Astrophysical Journal* **504**, pp. 492–499

[207] Nittler, L. R. (2005). Constraints on Heterogeneous Galactic Chemical Evolution from Meteoritic Stardust *The Astrophysical Journal* **618**, pp. 281–296

[208] Nittler, L. R., and Alexander, C. M. O'd. (1999). Can Stellar Dynamics Explain the Metallicity Distributions of Presolar Grains? *The Astrophysical Journal* **526**, pp. 249–256

[209] Nittler, L. R., and Alexander, C. M. O'd. (1999). Automatic Identification of Presolar Al- and Ti-rich Oxide Grains from Ordinary Chondrites *Lunar Planetary Science* **30**, abstract # 2041

[210] Nittler, L. R., and Alexander, C. M. O. (2003). Automated isotopic measurements of micron-sized dust: application to meteoritic presolar silicon carbide *Geochimica et Cosmochimica Acta* **67**, pp. 4961–4980

[211] Nittler, L. R., Alexander, C. M. O., Gao, X., Walker, R. M., and Zinner, E. (1994). Interstellar Oxide Grains from the Tieschitz Ordinary Chondrite *Nature* **370**, pp. 443–446

[212] Nittler, L. R., Alexander, C. M. O'd., Gao, X., Walker, R. M., and Zinner, E. (1997). Stellar Sapphires: The Properties and Origins of Presolar Al_2O_3 in Meteorites *The Astrophysical Journal* **483**, pp. 475–495

[213] Nittler, L. R., Alexander, C. M. O'd., Wang, J., and Gao, X. (1998). Meteoritic oxide grain from supernova found *Nature* **393**, pp. 222

[214] Nittler, L. R., Amari, S., Zinner, E., Woosley, S. E., and Lewis, R. S.

(1996). Extinct ^{44}Ti in Presolar Graphite and SiC: Proof of a Supernova Origin *Astrophysical Journal Letters* **462**, pp. L31–L34

[215] Nittler, L. R., Hoppe, P., Alexander, C. M. O'd., Amari, S., Eberhardt, P., Gao, X., Lewis, R. S., Strebel, R., Walker, R. M., and Zinner, E. (1995). Silicon Nitride from Supernovæ *Astrophysical Journal Letters* **453**, pp. L25–L28

[216] Nollett, K. M., Busso, M., and Wasserburg, G. J. (2003). Cool Bottom Processes on the Thermally Pulsing Asymptotic Giant Branch and the Isotopic Composition of Circumstellar Dust Grains *The Astrophysical Journal* **582**, pp. 1036–1058

[217] Ott, U. (1996). Interstellar Diamond Xenon and Timescales of Supernova Ejecta *Astrophysical Journal* **463**, pp. 344–348

[218] Ott, U., and Begemann, F. (1990a). Discovery of s-process barium in the Murchison meteorite *Astrophysical Journal Letters* **353**, pp. L57–L60.

[219] Ott, U., and Begemann, F. (1990b). S-Process Material in Murchison: Sr and More on Ba *Lunar Planetary Science* **21**, pp. 920

[220] Ott, U., and Begemann, F. (20020). Spallation recoil and age of presolar grains in meteorites *Meteoritics & Planetary Science* **35**, pp. 53–63.

[221] Owen, T., Mahaffy, P. R., Niemann, H. B., Atreya, S., and Wong, M. (2001). Protosolar Nitrogen *The Astrophysical Journal Letters* **553**, pp. L77–79

[222] Patzer, A. B. C.; Gauger, A., and Sedlmayr, E. (1998). Dust formation in stellar winds. VII. Kinetic nucleation theory for chemical non-equilibrium in the gas phase *Astronomy and Astrophysics* **337**, pp. 847–858

[223] Pellin, M. J., Calaway, W. F., Davis, A. M., Lewis, R. S., Amari, S., and Clayton, R. N. (2000). Toward Complete Isotopic Analysis of Individual Presolar Silicon Carbide Grains: C, N, Si, Sr, Zr, Mo, and Ba in Single Grains of Type X *Lunar and Planetary Science* **31**, abstract # 1917

[224] Podosek, F. A., Prombo, C. A., Amari, S., and Lewis, R. S. (2004). s-Process Sr Isotopic Compositions in Presolar SiC from the Murchison Meteorite *The Astrophysical Journal* **605**, pp. 960–965

[225] Prinzhofer, A., Papanastassiou, D. A., and Wasserburg, G. J. (1989). The presence of Sm-146 in the early solar system and implications for its nucleosynthesis *Astrophysical Journal* **344**, pp. L81–L84

[226] Prombo, C. A., Podosek, F. A., Amari, S., and Lewis, R. S. (1993). S-process BA isotopic compositions in presolar SiC from the Murchison meteorite *Astrophysical Journal* **410**, pp. 393–399

[227] Qian, Y.-Z., and Wasserburg, G. J. (2003). Stellar Sources for Heavy r-Process Nuclei *The Astrophysical Journal* **588**, pp. 1099–1109

[228] Raiteri, C. M., Gallino, R., Busso, M., Neuberger, D., and Käppeler, F. (1993). The Weak s-Component and Nucleosynthesis in Massive Stars *Astrophysical Journal* **419**, p.207–223

[229] Rauscher, T., and Thielemann, F.-K. (2000). Astrophysical Reaction Rates From Statistical Model Calculations *Atomic Data and Nuclear Data Tables* **75**, pp. 1–351

[230] Rayet, M., Arnould, M., Hashimoto, M., Prantzos, N., and Nomoto, K.

(1995). The p-process in Type II supernovæ *Astronomy and Astrophysics* **298**, pp. 517– 527

[231] Reifarth, R., Heil, M., Käppeler, F., Voss, F., Wisshak, K., Bečvář, F., Krtička, M., Gallino, R., and Nagai, Y. (2002). Stellar neutron capture cross sections of ^{128,129,130}Xe *Physical Review C* **66**, id. 064603

[232] Reifarth, R., Käppeler, F., Voss, F., Wisshak, K., Gallino, R., Pignatari, M., and Straniero, O. (2004). ^{128}Xe and ^{130}Xe: Testing He-Shell Burning in Asymptotic Giant Branch Stars *The Astrophysical Journal* **614**, pp. 363– 370

[233] Reynolds, J. H., and Turner, G. (1964). Rare gases in the chondrite Renazzo, *J. Geophys. Res.* **69**, pp. 3263–81

[234] Richter, S., Ott, U., and Begemann, F. (1992). s-Process Isotope Anomalies: Neodymium, Samarium, and a Bit More of Strontium *Lunar and Planetary Science Conference* **23**, pp. 1147–1148

[235] Richter, S., Ott, U., and Begemann, F. (1993). s-Process isotope abundance anomalies in meteoritic silicon carbide: new data in Proceedings of the 2nd International Symposium on Nuclear Astrophysics, F. Kaeppeler and K. Wisshak eds. *Bristol: IOP Publishing*, pp. 127–132

[236] Richter, S., Ott, U., and Begemann, F. (1994). s-Process isotope abundance anomalies in meteoritic silicon carbide: Data for Dysprosium *Meteoritics* **29**, pp. 522–523

[237] Richter, S., Ott, U., and Begemann, F. (1998). Tellurium in pre-solar diamonds as an indicator for rapid separation of supernova ejecta *Nature* **391**, pp. 261–263

[238] Romano, D., and Matteucci, F. (2003). Nova nucleosynthesis and Galactic evolution of the CNO isotopes *Monthly Notice of the Royal Astronomical Society* **342**, pp. 185–198

[239] Russell, S. R., Arden, J. W., and Pwlinger, C. T. (1996). A carbon and nitrogen isotope study of diamond from primitive chondrites *Meteoritics and Planetary Science* **31**, pp. 343–355

[240] Russell, S. S., Ott, U., Alexander, C. M. O'd., Zinner, E., Arden, J. W., and Pillinger, C. T. (1997). Presolar silicon carbide from the Indarch (EH4) meteorite: Comparison with SiC populations from other meteorite classes *Meteoritics* **32**, pp. 719–732

[241] Sackmann, I.-J., and Boothroyd, A. I. (1992). The creation of superrich lithium giants *Astrophysical Journal Letters* **392** pp. 71–74

[242] Sackmann, I.-J., and Boothroyd, A. I. (1999). Creation of ^{7}Li and Destruction of ^{3}He, ^{9}Be, ^{10}B, and ^{11}B in Low-Mass Red Giants, Due to Deep Circulation *The Astrophysical Journal* **510**, pp. 217–231

[243] Savina, M. R., Pellin, M. J., Tripa, C. E., Veryovkin, I. V., Calaway, W. F., and Davis, A. M. (2003). Analyzing individual presolar grains with CHARISMA *Geochimica et Cosmochimica Acta* **67**, pp. 3215–3225

[244] Savina, M. R., Pellin, M. J., Tripa, C. E., Davis, A. M., Lewis, R. S., and Amari, S. (2004). Excess p-process molybdenum and ruthenium in a presolar SiC grain *Nuclear Physics A* , pp. –

[245] Savina M. R., Davis A. M., Tripa C. E., Pellin M. J., Clayton R. N., Lewis

R. S., Amari S., Gallino R., and Lugaro M. (2003). Barium isotopes in individual presolar mainstream silicon carbide grains from the Murchison meteorite *Geochimica et Cosmochimica Acta* **67**, pp. 3201–3214

[246] Savina, M. R., Davis, A. M., Tripa, C. E., Pellin, M. J., Gallino, R., Lewis, R. S., and Amari, S. (2004). Extinct Technetium in Silicon Carbide Stardust Grains: Implications for Stellar Nucleosynthesis *Science* **303**, pp. 649–652

[247] Schatz, H., Aprahamian, A., Goerres, J., Wiescher, M., Rauscher, T., Rembges, J. F., Thielemann, F.-K., Pfeiffer, B., Moeller, P., Kratz, K.-L., Herndl, H., Brown, B. A., and Rebel, H. (1998). rp-Process Nucleosynthesis at Extreme Temperature and Density Conditions *Physics Reports* **294**,pp. 167–264

[248] Schramm, D. N., and Wasserburg, G. J. (1970). Nucleochronologies and the Mean Age of the Elements *Astrophysical Journal* **162**, pp. 57–70

[249] Schwarzschild, M., and Härm, R. (1965). Thermal Instability in Non-Degenerate Stars *Astrophysical Journal* **142**, pp. 855–867

[250] Sedlmayr, E., and Krüger, D. (1997). Formation of dust particles in cool stellar outflows in Astrophysical Implications of the Laboratory Study of Presolar Materials. Edited by Thomas J. Bernatowicz and Ernst Zinner *American Institute of Physics Conference proceedings 402*, pp. 307–315

[251] Sellwood, J. A., and Binney, J. J. (2002). Radial mixing in galactic discs *Monthly Notice of the Royal Astronomical Society* **336**, pp. 785–796

[252] Sharp, C. M., and Wasserburg, G. J. (1995). Molecular equilibria and condensation temperatures in carbon-rich gases *Geochimica et Cosmochimica Acta* **59**, pp. 1633–1652

[253] Siess, L., Goriely, S., and Langer, N. (2004). Nucleosynthesis of s-elements in rotating AGB stars *Astronomy and Astrophysics* **415**, pp. 1089–1097

[254] Smith, V. V., and Lambert, D. L. (1986). The chemical composition of red giants. II - Helium burning and the s-process in the MS and S stars *Astrophysical Journal* **311**, pp. 843–863

[255] Smith, V. V., and Lambert, D. L. (1990). The chemical composition of red giants. III - Further CNO isotopic and s-process abundances in thermally pulsing asymptotic giant branch stars *Astrophysical Journal Supplement Series* **72**, pp. 387–416

[256] Sneden, C., Cowan, J. J., Lawler, J. E., Ivans, I. I., Burles, S., Beers, T. C., Primas, F., Hill, V., Truran, J. W., Fuller, G. M., Pfeiffer, B., and Kratz, K.-L. The Extremely Metal-poor, Neutron Capture-rich Star CS 22892-052: A Comprehensive Abundance Analysis *The Astrophysical Journal* **591**, pp. 936–953

[257] Speck, A. K., Barlow, M. J., Sylvester, R. J., and Hofmeister, A. M. (2000). Dust features in the 10-mu m infrared spectra of oxygen-rich evolved stars *Astronomy and Astrophysics Supplement* **146**, pp. 437–464

[258] Speck, A. K., and Hofmeister, A. M. (2004). Processing of Presolar Grains around Post-Asymptotic Giant Branch Stars: Silicon Carbide as the Carrier of the 21 Micron Feature *The Astrophysical Journal* **600**, pp. 986–991

[259] Speck, A. K., Hofmeister, A. M., and Barlow, M. J. (1999). The SIC Problem: Astronomical and Meteoritic Evidence *Astrophysical Journal Letters* **513**, pp. L87–L90

[260] Srinivasan, B., and Anders, E. (1978). Noble gases in the Murchison meteorite - Possible relics of s-process nucleosynthesis, *Science* **201**, pp. 51-56

[261] Stadermann, F. J., Croat, T. K., and Bernatowicz, T. (2004). NanoSIMS Determination of Carbon and Oxygen Isotopic Compositions of Presolar Graphites from the Murchison Meteorite *Lunar and Planetary Science* **35**, abstract # 1758

[262] Stancliffe, R., Tout, C. A., and Pols, O. (2004). Deep Dredge-up in Intermediate-Mass Thermally Pulsing AGB stars *Monthly Notices of the Royal Astronomical Society*, in press

[263] Starrfield, S., Gehrz, R. D., and Truran, J. W. (1997). Dust Formation and Nucleosynthesis in the Nova Outburst, in Astrophysical Implications of the Laboratory Study of Presolar Materials. Edited by Thomas J. Bernatowicz and Ernst Zinner *American Institute of Physics Conference proceedings 402*, pp. 203–234

[264] Stone, J., Hutcheon, I. D., Epstein, S., and Wasserburg, G. J. (1991). Correlated SI isotope anomalies and large C-13 enrichments in a family of exotic SiC grains *Earth and Planetary Science Letters* **107**, pp. 570–581

[265] Straniero, O., Chieffi, A., Limongi, M., Busso, M., Gallino, R., and Arlandini, C. (1997). Evolution and Nucleosynthesis in Low-Mass Asymptotic Giant Branch Stars. I. Formation of Population I Carbon Stars *Astrophysical Journal* **478**, pp. 332–339

[266] Straniero, O., Gallino, R., Busso, M., Chieffi, A., Raiteri, C. M., Limongi, M., and Salaris, M. (1995). Radiative C-13 burning in asymptotic giant branch stars and s-processing *Astrophysical Journal Letters* **440**, pp. L85–L87

[267] Suess, H. E., and Urey, H. C. (1956). Abundances of the Elements. *Reviews of Modern Physics* **28**, pp. 53–74

[268] Sumiyoshi, K., Terasawa, M., Mathews, G. J., Kajino, T., Yamada, S., and Suzuki, H. (2001). r-Process in Prompt Supernova Explosions Revisited *The Astrophysical Journal* **562**, pp. 880–886

[269] Sweigart, A. V., and Mengel, J. G. (1979). Meridional circulation and CNO anomalies in red giant stars *Astrophysical Journal* **229**, pp. 624–641

[270] Tang M., and Edward, A. (1988). Isotopic anomalies of Ne, Xe, and C in meteorites. II. Interstellar diamond and SiC: Carriers of exotic noble gases *Geochimica et Cosmochimica Acta* **52**, pp. 1235–1244

[271] Tang, M., Anders, E., Hoppe, P., and Zinner, E. (1989). meteoritic silicon carbide and its stellar sources: implications for galactic chemical evolution *Nature* **339**, pp. 351–354

[272] Terasawa, M., Sumiyoshi, K., Kajino, T., Mathews, G. J., and Tanihata, I. (2001). New Nuclear Reaction Flow during r-Process Nucleosynthesis in Supernovæ: Critical Role of Light, Neutron-rich Nuclei *The Astrophysical Journal* **562**, pp. 470–479

[273] Terasawa, M., Sumiyoshi, K., Yamada, S., Suzuki, H., and Kajino, T. *r*-Process Nucleosynthesis in Neutrino-driven Winds from a Typical Neutron Star with $M = 1.4\,M_\odot$ *Astrophysical Journal Letters* **578**, pp. L137–L140

[274] Thielemann, F.-K., Nomoto, K., and Yokoi, K. (1986). Explosive nucle-

osynthesis in carbon deflagration models of Type I supernovæ *Astronomy and Astrophysics* **158**, pp. 17-33

[275] Thiemens, M. H. (1999). Mass-Independent Isotope Effects in Planetary Atmospheres and the Early Solar System *Science* **283**, pp. 341–345

[276] Thompson, T. A. (2003). Magnetic Protoneutron Star Winds and r-Process Nucleosynthesis *Astrophysical Journal Letters* **585**, pp. L33–L36

[277] Tielens, A. G. G. M. (1997). Deuterium and interstellar chemical processes in Astrophysical Implications of the Laboratory Study of Presolar Materials. Edited by Thomas J. Bernatowicz and E. Zinner *American Institute of Physics Conference proceedings 402*, pp. 523–344

[278] Timmes, F. X., and Clayton, D. D. (1996). Galactic Evolution of Silicon Isotopes: Application to Presolar SiC Grains from Meteorites *Astrophysical Journal* **472**, pp. 723–741

[279] Timmes, F. X., Woosley, S. E., and Weaver, T. A. (1995). Galactic chemical evolution: Hygrogen through zinc *Astrophysical Journal Supplement Series* **98**, pp. 617–658

[280] Toukan, K. A., and Kaeppeler, F. (1990). The stellar neutron capture cross sections of Zr-94 and Zr-96 *Astrophysical Journal* **348**, pp. 357–362

[281] Travaglio, C., Gallino, R., Amari, S., Zinner, E., Woosley, S., and Lewis, R. S. (1999). Low-Density Graphite Grains and Mixing in Type II Supernovæ *The Astrophysical Journal* **510**, pp. 325–354

[282] Treffers, R., and Cohen, M. (1974). High-resolution spectra of cool stars in the 10- and 20-micron regions *Astrophysical Journal* **188**, pp. 545–552

[283] Truran, J. W., and Iben, I., Jr. (1977). On s-process nucleosynthesis in thermally pulsing stars *Astrophysical Journal* **216**, pp. 797–810

[284] Ventura, P., and D'Antona, F. (2005). Full computation of massive AGB evolution. I. The large impact of convection on nucleosynthesis *Astronomy and Astrophysics* **431**, pp. 279–288

[285] Verchovsky, A. B., Fisenko, A. V., Semjonova, L. F., Wright, I. P., Lee, M. R., and Pillinger, C. T. (1998). C, N and Noble Gas Isotopes in Grain Size Separates of Presolar Diamonds from Eframovka *Science* **282**, pp. 1165–1168

[286] Verchovsky, A. B., Wright, I. P., and Pillinger, C. T. (2004). Astrophysical Significance of AGB Stellar Wind Energies Recorded in Meteoritic SiC Grains *The Astrophysical Journal* **607**, pp. 611–619

[287] Wallerstein, G. *et al* (1997). Synthesis of the elements in stars: forty years of progress *Reviews of Modern Physics* **69**, pp. 995–1084

[288] Ward, R. A., Newman, M. J., and Clayton, D. D. (1976). S-process Studies: Branching and the Time Scale *Astrophysical Journal Supplement* **31**, pp. 33–59

[289] Wasserburg, G. J., Boothroyd, A. I., and Sackmann, I.-J. (1995). Deep Circulation in Red Giant Stars: A Solution to the Carbon and Oxygen Isotope Puzzles? *Astrophysical Journal Letters* **447**, pp. L37–L40

[290] Wasserburg, G. J., Busso, M., Gallino, R., and Raiteri, C. M. (1994). Asymptotic Giant Branch stars as a source of short-lived radioactive nuclei in the solar nebula *Astrophysical Journal* **424**, pp. 412–428

[291] Waters, L. B. F. M et al.(1996). Mineralogy of oxygen-rich dust shells *Astronomy and Astrophysics Letters* **315**, pp. L361–L364

[292] Weizsäcker, V. C. F. V. (1937). Über elementumwandlungen im innern der sterne *Zeitschrift für Physik* **38**, pp. 176 – 199

[293] Wielen, R., Fuchs, B., and Dettbarn, C. (1996). On the birth-place of the Sun and the places of formation of other nearby stars *Astronomy and Astrophysics Letters* **314**, pp. 438–447

[294] Wooden, D. H. (1997). Observational evidence for mixing and dust condensation in core-collapse supernovæ in Astrophysical Implications of the Laboratory Study of Presolar Materials. Edited by Thomas J. Bernatowicz and Ernst Zinner *American Institute of Physics Conference proceedings 402*, pp. 317–376

[295] Woosley, S. E. (1997). Neutron-rich Nucleosynthesis in Carbon Deflagration Supernovæ *Astrophysical Journal* **476**, pp. 801–810

[296] Woosley, S. E., Arnett, W. D., and Clayton, D. D. (1973). The Explosive Burning of Oxygen and Silicon *Astrophysical Journal Supplement* **26**, pp. 231–312

[297] Woosley, S. E., Heger, A., and Weaver, T. A. (2002). The evolution and explosion of massive stars *Reviews of Modern Physics* **74**, pp. 1015–1071

[298] Woosley, S. E., and Weaver, T. A. (1995). The Evolution and Explosion of Massive Stars. II. Explosive Hydrodynamics and Nucleosynthesis *Astrophysical Journal Supplement* **101**, pp. 181–230

[299] Yoshida, T., and Hashimoto, M. (2004). Numerical Analyses of Isotopic Ratios of Presolar Grains from Supernovæ *The Astrophysical Journal* **606**, pp. 592–604.

[300] Yoshida, T., Umeda, H., and Nomoto, K. (2005). Silicon Isotopic Ratios of Presolar Grains from Supernovæ *The Astrophysical Journal*, to appear

[301] Zinner, E. (1998). Stellar Nucleosynthesis and the Isotopic Composition of Presolar Grains from Primitive Meteorites *Annual Review of Earth and Planetary Sciences* **26**, pp. 147–188

[302] Zinner, E. (2004). Presolar grains, in Treatise on Geochemistry. 1. Edited by K.K., Turekian, H.D. Holland, and A.D. Davis *Elsevier*, Oxford and San Diego, pp. 17–39

[303] Zinner, E., Amari, S., Gallino, R., and Lugaro, M. (2001). Evidence for a range of metallicities in the parent stars of presolar SiC grains *Nuclear Physics A* **688**, pp. 102–105

[304] Zinner, E., Amari, S., Guinness, R., Nguyen, A., Stadermann, F. J., Walker, R. M., and Lewis, R. S. (2003). Presolar spinel grains from the Murray and Murchison carbonaceous chondrites *Geochimica et Cosmochimica Acta* **67**, pp. 5083–5095

[305] Zinner, E., Amari, S., and Lewis, R. S. (1991). *s*-process Ba, Nd, and Sm in presolar SiC from the Murchison meteorite *Astrophysical Journal Letters* **382**, pp. L47–L50.

[306] Zinner, E., Ming, T., and Anders, E. (1987). Large isotopic anomalies of Si, C, N and noble gases in interstellar silicon carbide from the Murray meteorite *Nature* **330**, pp. 730–732

[307] Zinner, E., Nittler, L. R., Hoppe, P., Gallino, R., Straniero, O., and Alexander, C. M. O'd. (2005). Oxygen, magnesium and chromium siotopic ratios of presolar spinel grains *Geochimica et Cosmochimica Acta*, to appear

[308] Zinner, E., Amari, S., Jennings, C., Mertz, A., Nguyen, A., Nittler, L. R., Hoppe, P., Gallino, R., and Lugaro, M. (2005). Al and Ti isotopic ratios of presolar SiC grains of type Z *Lunar and Planetary Science* **36**, abstract # 1691

Index

α nuclei, 39, 42, 44, 165
ν process, 55, 56, 165
^{13}C pocket, 117, 122, 123, 128, 129,
    133, 135–138, 143
    formation, 119–121
pp chain, 29, 31, 97, 167

alpha-rich freeze out, 44, 54, 110, 165
Asymptotic Giant Branch (AGB), 15,
    17, 22, 37, 38, 49, 83–86, 93, 114,
    154, 155, 159–162, 165–167

C/O, 12, 17, 21, 22, 82, 83, 87, 92, 93,
    110, 111, 127, 143, 155, 156, 159
CAIs, 3, 5, 11
carbon star, 87, 88, 96, 99, 128, 140
    C(J), 109
    C(N), 83
condensation, 12, 13, 17, 21, 22, 63,
    113, 153, 170, 171
cool bottom processing, 20, 93, 160,
    166
cosmic rays, 4, 23, 97, 166
cycle, 32–35, 167
    CNO, 33, 88, 98, 117, 161
    MgAl, 35, 98
    NeNa, 35

degeneracy, 35, 37–39, 44, 54, 85, 88,
    136, 165, 166, 168, 171
diamond, 10–14, 23, 24, 46, 54, 57,
    59–62, 111, 151, 153, 154

dredge-up, 19, 86, 166
    first, 84, 88–93, 97, 111, 159–161,
        166
    second, 85, 161, 166
    third, 87, 92, 93, 95, 98, 108, 111,
        118–120, 122, 125, 127–129,
        140, 143, 162, 165, 166

electron microscope, 15, 16, 59, 62,
    63, 71
    SEM, 62, 63, 73, 74
    TEM, 63
entropy, 40, 41
extra mixing, 20, 91–93, 108, 111, 160

fractionation, 5, 18, 97, 167

Galactic Chemical Evolution (GCE),
    1, 7, 19, 93, 94, 102–104, 107, 112,
    135, 160–163, 167
graphite, 10–12, 16, 17, 20–22, 24, 36,
    44, 46, 59–62, 66, 70, 103, 110, 146,
    151, 153–156

hot bottom burning, 20, 87, 101, 116,
    160, 162, 167

internal gravity waves, 121
Interplanetary Dust Particles (IDPs),
    5, 14, 74, 167
interstellar medium (ISM), 1, 2, 5, 7,
    19, 23, 38, 43, 97, 103–105, 135,

167, 168
ion
  detector, 65
  imaging, 59, 71, 73, 74
  implantation, 13, 21, 94, 113, 144, 167, 171
  microprobe, 65, 71, 73, 74

magic number, 46, 48, 52, 152, 168
main sequence, 29, 83–85, 88, 97, 168, 170
mainstream line, 80, 81, 99–101, 103, 105
mass loss, 6, 22, 35, 87, 129
mass spectrometer
  RIMS, 68–70, 107, 109, 111, 128, 133, 136, 137, 140, 141
  SIMS, 65, 66, 68, 69, 75
  TIMS, 68, 132, 133
meteorite
  carbonaceous, 8
  chondritic, 12, 26, 77, 97
  Murchison, 10–13, 15, 16, 23, 24, 60, 70, 77, 155, 168
  primitive, 5, 8, 23, 83, 157, 169
molecular cloud, 5, 7, 104, 168

NanoSIMS, 14, 17, 67–69, 74, 77, 107, 155
Ne-E
  (H), 8, 9, 16, 94, 101
  (L), 8, 10, 61, 154
neutron
  burst, 55, 109, 111, 151, 153, 154
  density, 45, 46, 49–51, 53, 114, 116, 122, 125, 138–140, 146, 147, 149, 170
  excess, 42, 44, 56, 58
  exposure, 50, 52, 122–124, 127, 132, 133, 135, 138, 139, 149
  flux, 46, 48, 50, 98, 111, 122–124, 129, 132, 136–138, 153
  source, 115–118, 122, 123, 125, 137, 148
  star, 43, 54–57, 165, 168

noble gas, 8, 13, 17, 21, 48, 65, 94, 97, 113, 142, 145, 154, 168
nova, 7, 22, 35, 36, 81, 93, 109, 154, 155, 161, 168, 170
nuclear
  force, 28
  reaction rate, 36, 57, 140
  statistical equilibrium, 53, 56
nucleosynthesis
  primary, 32, 54, 55, 93, 103, 107, 135, 161, 169
  secondary, 32, 35, 38, 39, 42, 54, 93, 103, 107, 161, 169
nuclide chart, 34, 46, 47, 49, 53, 58, 147, 152, 169

overshoot, 120, 124, 125, 140, 169
oxide grains, 11, 17, 24, 35, 71–73, 151, 157, 159, 161
  corundum, 11, 12, 17, 59, 73, 74, 157–160, 162
  hibonite, 17, 74, 157
  spinel, 12, 17, 60, 62, 73, 157, 162

peak, in the solar system abundances, 26, 27, 39, 45, 48
  r process, 48, 55
  s process, 48
  Fe, 27, 28, 43, 44, 53, 167
photodisintegration, 39, 40, 53, 169

red giant, 4, 6, 14, 15, 17, 19–22, 25, 28, 31, 35, 36, 49, 51, 82–84, 88, 91–93, 154, 155, 159, 160, 162, 166, 170
  chemically peculiar, 49, 82, 115

Schwarzschild criterion, 119, 120, 167
semiconvection, 119, 170
silicate grains, 11, 14, 73, 157, 159, 162
silicon carbide (SiC), 9, 11, 12, 14, 15, 17, 21–24, 31, 35, 36, 46, 49, 51, 59–63, 66, 69–71, 73
  age, 97
  mainstream, 80–83, 88, 91, 92,

98–102, 104, 105, 107, 108,
  112–114, 154, 160
SiC-A+B, 57, 80, 105, 106, 109
SiC-X, 17, 20, 22, 44, 46, 55, 72,
  73, 109, 110, 155, 156
SiC-Y, 105, 106, 108
SiC-Z, 80, 107, 108, 135
silicon nitride, 12, 17, 44, 73, 110, 155
solar system
  abundances, 1, 39, 45, 56, 116
  formation, 3–5, 23, 169
spallation, 23, 28, 55, 97, 165, 170,
  171
spectroscopic
  classification, 82
  observations, 1, 7, 91, 105, 114,
    115, 136, 159
stellar
  orbital diffusion, 104
  rotation, 20, 115, 122, 136, 137
  winds, 4, 22, 23, 87, 88, 97, 144,
    145, 147, 165, 171
supernova, 4, 7, 13, 17, 20, 22, 23, 28,
  37, 40, 43, 44, 57, 82, 104, 105, 107,
  110, 111, 154, 155, 166, 170, 171

type Ia (SNIa), 44, 57, 102, 103,
  107, 110, 154, 171
type II (SNII), 20, 43, 44, 53–55,
  57, 73, 82, 93, 102–104,
  109–111, 153–158, 161, 165,
  171

technetium, 25, 82
thermal pulse, 85, 87, 98, 109,
  115–117, 121, 122, 124, 125, 138,
  148
three-isotope plot, 94–96, 99–101,
  127, 128, 130, 133, 134, 137–139,
  142, 143, 148, 158, 159, 171

white dwarf, 35, 38, 44, 54, 56, 88,
  109, 154, 166, 168, 171
Wolf-Rayet stars, 154, 171

Xe
  H, 54, 111, 151, 153, 162
  HL, 8, 9, 13, 57, 58, 151, 154
  L, 154
  S, 8, 9, 15, 45, 58, 82, 114, 142, 144

Printed in the United States
By Bookmasters